Essential Molecular Biology
Volume One

University of Hertfordshire

Learning and Information Services

The Practical Approach Series

Related **Practical Approach** Series Titles

Human Cytogenetics: Malignancy and Acquired Abnormalities 2

Human Cytogenetics: Constitutional Analysis*

Animal Cell Culture 3/e

RNA Viruses

Differential Display

Mouse Genetics and Transgenics

DNA Viruses

Gene Targeting 2/e

Protein Phosphorylation 2/e

Crystallization of Nucleic Acids and Proteins

DNA Microarray Technology

Post-Translational Modification

Protein Expression

Chromosome Structural Analysis

Gel Electrophoresis of Proteins 3/e

In Situ Hybridization 2/e

Chromatin

Mutation Detection

Molecular Genetic Analysis of Populations 2/e

PCR 3: PCR In Situ Hybridization

Antisense Technology

Genome Mapping

Protein Function 2/e

Protein Structure 2/e

DNA and Protein Sequence Analysis

Protein Structure Prediction

Antibody Engineering

DNA Cloning 4: Mammalian Systems

DNA Cloning 3: Complex Genomes

Gene Probes 2

Gene Probes 1

Pulsed Field Gel Electrophoresis

Non-isotopic Methods in Molecular Biology

PCR 2

DNA Cloning 2: Expression Systems

DNA Cloning 1: Core Techniques

Molecular Genetics of Yeast

RNA Processing Volume II

RNA Processing Volume I

PCR 1

Plant Molecular Biology

* indicates a forthcoming title

Please see the **Practical Approach** series website at
http://www.oup.co.uk/pas

for full contents lists of all Practical Approach titles.

Essential Molecular Biology

Volume One

Second Edition

A Practical Approach

Edited by

T. A. Brown

Department of Biomolecular Sciences
UMIST, Manchester, U.K.

OXFORD
UNIVERSITY PRESS

OXFORD

UNIVERSITY PRESS

Great Clarendon Street, Oxford OX2 6DP

Oxford University Press is a department of the University of Oxford.
It furthers the University's objective of excellence in research,
scholarship, and education by publishing worldwide in

Oxford New York

Athens Auckland Bangkok Bogotá Buenos Aires Calcutta Cape Town
Chennai Dar es Salaam Delhi Florence Hong Kong Istanbul Karachi
Kuala Lumpur Madrid Melbourne Mexico City Mumbai Nairobi Paris
São Paulo Singapore Taipei Tokyo Toronto Warsaw

with associated companies in Berlin Ibadan

Oxford is a registered trade mark of Oxford University Press in the UK
and in certain other countries

Published in the United States by Oxford University Press Inc., New York

British Library Cataloguing in Publication Data
Data available

Library of Congress Cataloguing in Publication Data

1 3 5 7 9 10 8 6 4 2

ISBN 0 19 963643 5 (Hbk.)
ISBN 0 19 963642 7 (Pbk.)

Typeset in Swift by Footnote Graphics, Warminster, Wilts
Printed in Great Britain on acid-free paper
by The Bath Press, Avon

Preface to the first edition

There are now a number of molecular biology manuals on the market and the editor of an entirely new one has a duty to explain why his contribution should be needed. My answer is that although there are some excellent handbooks for researchers who already know the basic principles of gene cloning there are very few that cater for the absolute beginner. Unfortunately, everyone is a beginner at some stage in their careers and, even in an established molecular biology laboratory, the new research student can spend a substantial amount of time not really understanding what is going on. For the experienced biologist expert in a discipline other than molecular biology, and perhaps without direct access to a tame gene cloner, guidance on how to introduce recombinant DNA techniques into their own research programme can be very difficult to obtain. For several years I have run a basic gene cloning course at UMIST and I have continually been impressed by the number of biochemists, botanists, geneticists, cell biologists, medics and others who want to learn how to clone and study genes.

The contributors to *Essential Molecular Biology: A Practical Approach* were asked to write accounts that combine solid practical information with sufficient background material to ensure that the novice can understand how a technique works, what it achieves, and how to make modifications to suit personal requirements. Where appropriate, the reader is also given advice on more advanced or specialized techniques. In all cases the authors have responded to the challenge and produced chapters that make concessions to the beginner without jeopardizing scientific content or practical value. I hope that the result is a handbook that will guide newcomers into molecular biology research.

The book is split into two parts. Volume I deals with the fundamental techniques needed to carry out DNA cloning experiments. The emphasis is on coming to grips with the necessary practical skills and understanding the background in sufficient detail to be able to adjust to circumstances as the project progresses. In Volume II, procedures for preparing gene libraries and identifying genes are described, together with methods for studying the structure of a cloned gene and the way it is expressed in the cell. It is assumed that the basics from Volume I are now in place, but the procedures are still described in the same down-to-earth fashion, with protocols complemented by background information and troubleshooting hints.

I must thank a number of people for their help with this book. First, I am grateful to the authors who provided the manuscripts, more or less on time, and were prepared in many cases to make revisions according to my requests. I would especially like to thank Paul Towner for stepping in at the last minute after I had been let down with one chapter. The participants and assistants on the recent UMIST gene cloning courses helped me formulate the contents of the book and my colleague Paul Sims provided valuable advice. The Series Editors and publishers made encouraging noises when the going got tough, and sorted out a number of problems for me. I am also grateful to my research students for giving their opinion on the practical contents of the book. Finally, I would like to thank my wife Keri for proving that even an archaeologist can learn how to clone genes.

1990   T.A.B.

Preface to the Second Edition

The primary objective of *Essential Molecular Biology: A Practical Approach*, that the book should be accessible and informative to researchers encountering molecular biology techniques for the first time, is as valuable today as it was when the first edition was written 10 years ago. If anything, the need for a basic DNA manual has increased as molecular biology techniques have become more and more mainstream in disparate areas of research. Gene cloning, sequencing and PCR are now central to those various research fields grouped together under 'molecular life sciences' and an ever-increasing number of new research students enter these fields every year. Equally impressive has been the spread of DNA techniques into areas of research such as molecular ecology and biomolecular archaeology, where the power of DNA analysis is contributing to scientific knowledge in disciplines that just a few years ago either did not exist at all or had little use for clones and PCR products. When planning the second edition of *Essential Molecular Biology: A Practical Approach* I therefore determined from the very start that the objective of the book should remain the same: 'to combine solid practical information with sufficient background material to ensure that the novice can understand how a technique works, what it achieves, and how to make modifications to suit personal requirements'.

Since the first edition was published there has been a dramatic expansion in the range and sophistication of molecular biology techniques, even in those basic techniques that are of greatest importance to the newcomer. To address these changes, every chapter has been revised and updated as appropriate, with some chapters completely rewritten. This is particularly true for the more advanced techniques described in Volume II but also applies to the very basic procedures covered in Volume I, the advances over the last 10 years even involving such fundamental techniques as DNA purification and agarose gel electrophoresis. Despite these changes, the two volumes are still organized according to the rationale of the first edition, with the procedures for DNA and RNA manipulation (purification, electrophoresis and the construction and cloning of recombinant molecules) in Volume I and those for isolating and studying individual genes (preparation and screening of libraries, polymerase chain reactions, DNA sequencing and studying gene expression) in Volume II.

PREFACE TO THE SECOND EDITION

An edited volume is only as good as the chapters it contains and I therefore wish to thank all the contributors for taking such great care to address the objective of *Essential Molecular Biology* by making their chapters lucid and informative. I also thank Liz Owen and Lisa Blake of Oxford University Press for ensuring that the chapters became a book rather than languishing in my computer, and my wife Keri Brown for helping me find the time to organize and assemble these two volumes.

2000   T.A.B.

Contents

CONTENTS

CONTENTS

Protocol list

Abbreviations

A_{260}	absorption at 260 nm
BAP	bacterial alkaline phosphatase
Bis	bis(hydroxymethyl)aminomethane
bp	base pairs
BPB	bromophenol blue
CCD	charge-coupled device
cDNA	copy DNA
CIP	calf intestinal alkaline phosphatase
dATP	2'-deoxyadenosine 5'-triphosphate
dCTP	2'-deoxycytidine 5'-triphosphate
DEPC	diethyl pyrocarbonate
dGTP	2'-deoxyguanosine 5'-triphosphate
DMSO	dimethyl sulphoxide
DNA	deoxyribonucleic acid
DNase	deoxyribonuclease
dNTP	deoxyribonucleotide
DOC	sodium deoxycholate
DTT	dithiothreitol
dTTP	2'-deoxythymidine 5'-triphosphate
EDTA	ethylenediaminetetra-acetic acid
EEO	electroendosmosis
EGTA	ethyleneglycolbis(aminoethyl)tetra-acetic acid
GST	glutathione S-transferase
IAA	indole acrylic acid
IPTG	isopropyl-β-D-thiogalactopyranoside
kb	kilobase pairs
kDa	kilodaltons
LB	Luria–Bertani (media)
MCS	multi-cloning sequence
MES	2-morpholinoethanesulphonic acid monohydrate
MOPS	3-morpholinopropanesulphonic acid
mRNA	messenger RNA
nt	nucleotides

ABBREVIATIONS

NTA	nitrilotriacetic acid
NTP	nucleotide triphosphate
OD	optical density
PBS	phosphate-buffered saline
PCR	polymerase chain reaction
PEG	polyethylene glycol
PFGE	pulsed-field gel electrophoresis
PIPES	piperazine-1,4-bis(2-ethanesulphonic acid)
RBS	ribosome binding sequence
RF	replicative form
RNA	ribonucleic acid
RNase	ribonuclease
r.p.m.	revolutions per minute
rRNA	ribosomal RNA
SDS	sodium dodecyl sulphate
TdT	terminal deoxynucleotidyl transferase
TLC	thin-layer chromatography
Tris	tris(hydroxymethyl)aminomethane
tRNA	transfer RNA
UV	ultraviolet
XC	xylene cyanol
X-gal	5-bromo-4-chloro-3-indolyl-β-D-galactopyranoside

Chapter 1
Getting started in molecular biology

T. A. Brown

Department of Biomolecular Sciences, University of Manchester Institute of Science and Technology, Manchester M60 1QD, UK

1 Introduction

In several respects molecular biology is not a user-friendly discipline. The newcomer is faced with a bewildering array of techniques, some unhelpful jargon, and the impression that only the complicated experiments are worth doing. In fact, most of the apparent problems are illusory and a series of relatively straightforward experiments will allow any competent research worker to obtain information on the structure and mode of expression of a gene. This is true not only for scientists in established molecular biology laboratories but also for researchers from other areas wishing to apply DNA analysis techniques in their own projects.

The chapters which follow describe the procedures for purifying DNA and RNA, constructing recombinant DNA molecules and obtaining clones (Volume I), and for preparing gene libraries, identifying genes, carrying out PCRs, and studying the structure and expression of genes (Volume II). But before diving into the laboratory, the reader should first spend some time thinking about four general issues that underlie successful research in molecular biology. These are:

- The practical requirements: the experimental skills and the equipment needed for research with DNA.

- The health hazards inherent in molecular biology and the safety procedures that must be adopted.

- The research strategies that are followed when studying DNA.

- The need to plan a project that will be informative.

This introductory chapter covers each of these four topics in turn.

2 Practical requirements for molecular biology research

2.1 Experimental skills

The newcomer to molecular biology should have no concerns about the man-ipulative skills needed to study DNA. Any researcher who has purified an active enzyme, separated chromosomes by flow cytometry or micromanipulated yeast asci will have few problems with the procedures described in this book. Many techniques in molecular biology require nothing more demanding than the ability to pipette microlitre amounts of solutions from one tube to another. The trick is knowing what to pipette and when to pipette it.

Practical molecular biology might not be overly challenging but it would be a mistake to believe that sloppy work will lead to good results. Accuracy is obviously important but so is scrupulous cleanliness to avoid contamination of bacterial cultures, degradation of nucleic acid samples, and inactivation of en-zymes. The following are the key aspects of laboratory practice for a successful molecular biologist:

(1) Learn how to pipette down to 1 μl accurately and reproducibly

(2) Develop a steady hand for loading samples on to gels and for other man-ipulations (give up caffeine if it makes your hand shake)

(3) Do not allow stocks of enzymes to warm up to room temperature: keep in the freezer and take out only for as long as it takes to remove an aliquot

(4) Autoclave plasticware (pipette tips, microfuge tubes, etc.) and keep sterile before use; do not handle pipette tips

(5) Always use double-distilled water; make sure the glassware cleaning pro-gramme includes a final rinse in double-distilled water

(6) Get into the habit of wearing disposable gloves at all times

2.2 Equipment

Much of the equipment and other facilities needed for molecular biology re-search will already be present in a well equipped biology laboratory, but if you are moving into molecular biology from a different area then you will almost certainly need to obtain some specialist items. The lists that follow are intended for guidance and do not cover every possible experiment that you may wish to do. The detailed applications of these pieces of equipment are described in later chapters.

2.2.1 General facilities

You will require access to the following general facilities:

• Containment facilities that the meet the requirements of the genetic man-ipulation regulations that your laboratory is subject to (see Section 3.1)

- Microbiological facilities:
  - autoclaves for sterilizing media
  - clean areas for molecular biology work (laminar flow cabinets are useful but not essential)
- 37 °C room. Incubators can be used as an alternative but it must be possible to incubate Petri dishes as well as cultures in test-tubes and flasks from 5 ml to 2 l
- Glassware cleaning facility
- Darkroom facilities:
  - red-light darkroom for autoradiography, including tanks and accessories for developing X-ray films
  - short-wavelength (302 or 366 nm) UV transilluminator
  - CCD detector or Polaroid camera for recording agarose gels
- Cold room
- Sub-zero storage facilities:
  - −80 °C freezer
  - liquid nitrogen storage vessels
- Radiochemical handling and disposal facilities, primarily for ^{35}S and ^{32}P
- Wet-ice machine and dry-ice storage

2.2.2 Departmental equipment

These items can be shared by a number of groups:

- Ultracentrifuge capable of 70 000 r.p.m., with at least one fixed-angle and one swing-out rotor plus accessories. Density gradient centrifugation can take 48 h so one machine is rarely sufficient
- Refrigerated centrifuge capable of 25 000 r.p.m., plus rotors for 50–500 ml buckets
- UV spectrophotometer
- Liquid scintillation counter
- Sonicator with a variety of probes
- Four-point balance
- Platform incubators for 45 °C, 55 °C and 65 °C
- Vacuum drier (a freeze-drier is not essential)
- Vacuum oven
- 100 °C oven
- Automated DNA-sequencing machine (desirable but not essential)
- Oligonucleotide synthesis machine. This is optional: commercial synthesis of oligonucleotides is cheap and rapid so an in-house facility is economical only if demand for oligonucleotides is high

2.2.3 Laboratory equipment

Each group carrying out molecular biology experiments will require these items, possibly in multiple numbers:

- Bench-top centrifuge, capable of 5000 r.p.m. but not refrigerated
- 200 V power supply for agarose gels, electroelution, etc.
- Horizontal gel apparatus (e.g. 10×10 cm, 15×10 cm, and 15×15 cm)
- Microfuge
- Automatic pipettes (e.g. Gilson P20, P200, P1000); each researcher will need a set
- Water baths (non-shaking): ideally three, one set at 37 °C, one at 65 °C and one varied for requirements
- Two-point top pan balance
- Vortex mixer
- Magnetic stirrer
- pH meter
- Geiger counter or other type of radioactivity monitoring device
- Microwave oven for melting agar and preparing agarose
- UV crosslinker
- Thermal cycler for PCR
- Hybridization oven or, as a cheap (and nasty) alternative, a bag sealer
- Cassettes and intensifying screens for autoradiography
- Light-box for viewing autoradiographs
- Double-distilled water supply
- Refrigerator
- -20 °C freezer
- Fume cupboard
- Computer: for virtually all molecular biology applications a networked PC or Macintosh is adequate

If DNA sequencing is being performed manually then you will also need:

- Vertical gel apparatus (e.g. 21×40 cm)
- 3000 V power supply
- Gel drier

For some types of hybridization analysis the following will be needed:

- Slot/dot blotter
- Electrotransfer or vacuum blotting unit

2.3 A word on kits

Many commercial suppliers offer a range of molecular biology kits that provide all the reagents and other materials needed for a particular experiment, together with protocols and sample DNA. So, for example, you can buy DNA labelling kits, cDNA synthesis kits, plus many others. These kits have become central to molecular biology research and several of the protocols described in this book make use of them. This is because kits usually provide reasonably good results and in most cases these results are easier to obtain than by the traditional method of purchasing the materials separately, making up one's own buffers, and suchlike. Cost is also a factor because kits are usually cheaper than the sum of the individual materials they contain.

Kits also have their negative side. The most significant of these is that although kits enable experiments to be carried out quickly they rarely provide the researcher with any flexibility with regard to the design of a procedure. The traditional route, although slower, will usually give better results in the long term, as experience gained along the way can be used to modify steps and buffer compositions to suit the particular DNA molecules being studied. A twofold improvement in a procedure can have a major effect on the outcome of the project. A second criticism of kits is their insidious property of substituting for laboratory skills. You will never understand a procedure in depth if all you know about a key reagent is that it is contained in the blue tube and you need to use 2 µl of it. Without understanding a procedure you will not be able to use it to its full advantage, know what to do when things go wrong, or appreciate when a result is artefactual or aberrant. The best advice is to learn and understand a technique first, then when you have it working see if a commercial kit is just as efficient and if so use the kit for repetitive experiments. If you are a research student, do not get the idea that using kits is the same as being a molecular biologist.

3 Health hazards and safety procedures

Safety is an individual responsibility and you must ensure that you are aware of the health hazards and have identified the appropriate safety requirements before embarking on molecular biology experiments. You should consult your departmental and institutional officers about the necessary precautions to take and be aware of any national health and safety regulations pertaining to the procedures that you will carry out. The following sections give an overview of molecular biology safety procedures and should be supplemented by reading the safety advice given in other chapters of this book and by referring to specialist publications.

3.1 Microbiological safety

Most molecular biology experiments are carried out with attenuated strains of *Escherichia coli* that are unable to survive outside of the test-tube. Nevertheless, precautions to avoid contamination, as described in Chapter 2, Section 1.3, should be observed whenever bacteria are being handled. (For detailed information, consult refs. 1–8.)

As well as general precautions to avoid infection with microorganisms, molecular biologists must also follow the additional precautions necessary when handling genetically manipulated bacteria. In most countries, these procedures are laid down in law (e.g. refs. 9 and 10) and a license or other form of approval may be required before experiments can be carried out.

3.2 Radiochemicals

Radiochemicals labelled with ^{32}P or ^{35}S are often used in molecular biology experiments. These are not the most dangerous radionuclides in biological research but they are used in large amounts in DNA analyses. Your national and local regulations (e.g. refs. 11–13) for handling and disposal of radiochemicals should be followed at all times. Various publications give general guidelines for the safe handling of radiochemicals (e.g. refs. 14–19) but the clearest and most succinct advice for molecular biologists is contained in the radiochemicals catalogue of Amersham International (20) and can be summarized as follows:

- Understand the nature of the hazard and get trained in the safety procedures before beginning work with radiochemicals
- Plan ahead so that the minimum amount of time is spent handling radiochemicals
- Keep the maximum possible distance between yourself and sources of radiation
- Use shielding that is appropriate for the type of radiation emitted by the source being handled
- Keep radioactive materials in defined work areas
- Wear appropriate clothing and dosimeters to monitor exposure
- Monitor the work area frequently and follow appropriate procedures if a spill occurs
- Keep waste accumulation as low as possible
- At the end of the experiment, monitor yourself, wash thoroughly, and monitor again

3.3 Chemical hazards

Many hazardous chemicals are used in molecular biology research. The container in which a hazardous chemical is supplied will carry labelling that describes the nature of the risk and the safety procedures that should be adopted. Most suppliers use a labelling system that combines an easily recognized symbol that indicates the general nature of the hazard (e.g. 'irritant') with a series of code numbers that provide specific risk and safety information (e.g. 'irritating to eyes and respiratory system'). The most comprehensive coding system, used by many companies, is the European Commission Risk and Safety Phrases. These phrases are defined in the suppliers' catalogues and safety literature (21–24) and there are also many books that give more general advice on the safe handling of hazardous chemicals (1, 25–37).

A detailed list of chemicals used in molecular biology research, together with the appropriate health and safety advice for each one, can be found in ref. 1. The hazards fall into three main groups:

- *Organic solvents.* The most dangerous solvent used in molecular biology is phenol. It is no longer necessary to undertake the highly hazardous operation of phenol distillation, needed if solid phenol is purchased, because many companies now market phenol that does not need redistilling. Always wear gloves when carrying out phenol extractions and make sure that the gloves actually provide a barrier to phenol: some lightweight disposable gloves do not.

- *Mutagens and carcinogens.* Most chemicals that bind to DNA in the test-tube are mutagenic and/or carcinogenic. An example is ethidium bromide, which is used in large amounts to detect DNA in agarose gels. Some manuals recommend adding ethidium bromide to the electrophoresis buffer, but this should be avoided. It does not provide any major advantage over post-staining (unless you want to follow the progress of the electrophoresis) but it does increase the chance of spillage. Check that your local regulations allow ethidium bromide solutions to be poured down the sink and even if this is permitted consider using a decontamination procedure (*Appendix 2; Protocol 5*) (38–40) before disposal. Remember that gels stained with ethidium bromide are dangerous and should not be placed in the general laboratory waste.

- *Toxic chemicals.* The most dangerous toxic chemical used in molecular biology is acrylamide which is lethal if swallowed and can also exert toxic effects by contact with the skin. The hazard is extreme with solid acrylamide monomer because the powder can form aerial contamination when being weighed out. For this reason, solid acrylamide should never be purchased and instead premixed acrylamide solutions should always be used.

3.4 Ultraviolet radiation

The most dangerous piece of equipment used by molecular biologists is the UV transilluminator used to view agarose gels that have been stained with ethidium bromide. Standard laboratory models have recommended exposure levels of 4.5 s or less at 30 cm (12 in). Unprotected exposure of skin can lead to severe burns and anyone foolish enough to look directly at the UV source without eye protection risks loss of sight. Always use a suitable screen when observing agarose gels and wear face and eye protection when the screen is removed for photography. Be particularly careful that the wrists and neck are protected when you are handling a gel on a transilluminator.

3.5 High-voltage electricity

When working in a laboratory it is very easy to concentrate on the special hazards of the research environment and to forget about the general hazards of daily life. The electricity used to run research equipment is equally as dangerous as that in your home—more so in some cases because higher voltages are involved. The power supplies used to run electrophoresis gels can kill. Commercial gel apparatuses are designed so that they cannot be disassembled before the power supply is disconnected. Home-made equipment should incorporate similar safeguards and broken gel tanks should never be used. Leaky gel apparatus is potentially lethal.

4 Research strategies for molecular biology

Whatever the long-term aim of a molecular biology project, the initial objective is usually to isolate a single gene or other specified DNA segment from the organism being studied. This is the essential preliminary step to DNA sequencing, to examination of the expression profile of a gene, or for purification of the protein coded by a gene. This initial objective can be achieved in either of two ways:

(1) By DNA cloning, which involves insertion of the desired DNA molecule into a cloning vector, followed by replication of the cloning vector in a culture of *E. coli* bacteria.

(2) By PCR, which involves enzymatic amplification of the desired DNA fragment.

Before beginning a project it is clearly important to decide which of these approaches will be used. They are not mutually exclusive: if cloning is adopted then PCR might be used at a later stage to check that the correct gene has been isolated, and if the primary approach is PCR then cloning might still be involved when material is prepared for DNA sequencing. However, a decision must be made about whether to base the experimental strategy on cloning or PCR. This section describes cloning and PCR in outline, gives advice regarding the choice between the two approaches, and explains how the basic techniques described in Volume I are used in cloning and in PCR. Volume II, Chapter 1 continues this theme by explaining the strategies used once the desired gene has been isolated.

4.1 Gene cloning in outline

The term 'gene cloning' is generally used, even though the technique is applicable to isolation of any DNA segment, not just those pieces containing one or more genes. The basic steps in a gene cloning experiment are shown in *Figure 1*. They are as follows (41):

(1) DNA is prepared from the organism being studied and fragments inserted into vector molecules to produce chimeras called recombinant DNA molecules.

(2) The recombinant DNA molecules are introduced into host cells. When cloning is used for gene isolation the cells are usually *E. coli* bacteria, but for specialized applications genes can be cloned into yeast or fungal cells, or into cells within intact animals and plants.

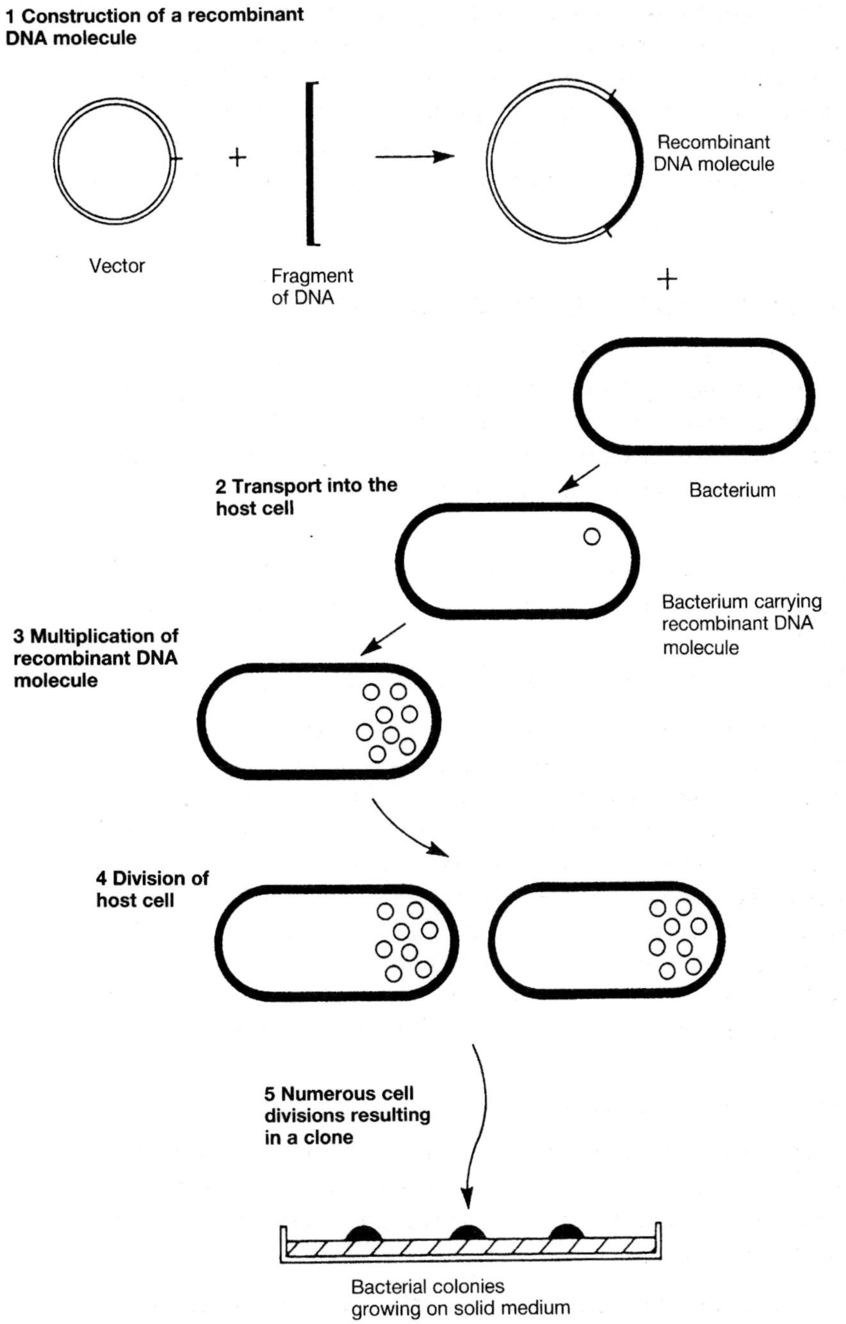

1 Construction of a recombinant DNA molecule

Vector

Fragment of DNA

Recombinant DNA molecule

2 Transport into the host cell

Bacterium

Bacterium carrying recombinant DNA molecule

3 Multiplication of recombinant DNA molecule

4 Division of host cell

5 Numerous cell divisions resulting in a clone

Bacterial colonies growing on solid medium

Figure 1 The basic steps in a gene cloning experiment. Reproduced, with permission, from Brown, T. A. (1995), *Gene cloning: an introduction*, 3rd edn. Chapman and Hall, London.

(3) The recombinant DNA molecules replicate within the host cells.

(4) When the host cell divides, copies of the recombinant DNA molecules are passed on to the progeny.

(5) Continued replication of the host cells results in clones, each of which consists of identical cells all containing copies of a single recombinant DNA molecule.

This series of manipulations results in a clone library, comprising many different clones, each carrying a different segment of the original DNA. The next step is therefore to identify the clone that contains the gene of interest, using one of the strategies described in Volume II, Chapter 1.

4.2 PCR in outline

PCR is a very different approach to isolation of a DNA segment. Rather than a lengthy series of manipulations involving living cells, PCR is a test-tube reaction that is carried out simply by mixing together the appropriate reagents and incubating them in a thermal cycler, a piece of equipment that enables the incubation temperature to be varied over time in a pre-programmed manner. The basic steps in a PCR experiment (*Figure 2*) are as follows (41):

(1) DNA is prepared from the organism being studied and denatured by heating to 94°C.

(2) A pair of oligonucleotides is added to the DNA; the sequences of these oligonucleotides enable them to anneal either side of the gene or other DNA segment that is to be isolated, and the mixture is cooled to 50–60°C so that these oligonucleotides attach to their target sites.

(3) A thermostable DNA polymerase is added together with a supply of deoxyribonucleotides and the mixture is heated to the optimal temperature for DNA synthesis, 74°C if *Taq* DNA polymerase (the DNA polymerase I enzyme from the thermotolerant bacterium *Thermus aquaticus*) is used.

(4) The cycle of denaturation-annealing-extension is repeated 25–30 times, with the number of newly-synthesized DNA molecules doubling during each cycle. This exponential amplification results in synthesis of a large number of copies of the DNA sequence flanked by the pair of oligonucleotides.

Fuller details of the events occurring during PCR are given in Volume II, Chapter 7.

4.3 The choice between cloning and PCR

Cloning and PCR both achieve the same end: they provide a sample of a single, short DNA molecule derived from within a lengthy starting molecule. The clearest difference between the two techniques is the time required to complete the procedure. Cloning is time-consuming, especially considering that the steps illustrated in *Figure 1* comprise just the first stage of a cloning experiment; further lengthy manipulations are often needed to identify the clone of interest from the resulting library. A competent researcher would expect to spend 2–3

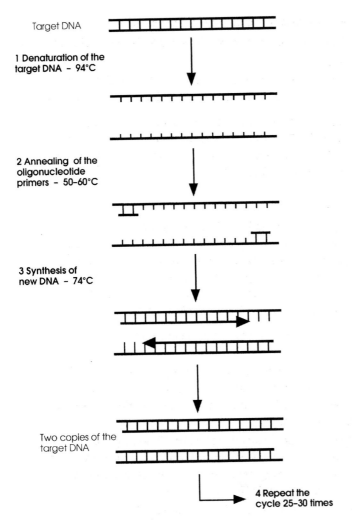

Target DNA

1 Denaturation of the
target DNA – 94°C

2 Annealing of the
oligonucleotide
primers – 50–60°C

3 Synthesis of
new DNA – 74°C

Two copies of the
target DNA

4 Repeat the
cycle 25–30 times

Figure 2 An outline of PCR.

months isolating a gene by cloning, but could achieve the same result in a single day by PCR. In view of the ease and rapidity of the PCR approach, it might seem unlikely that cloning would ever be the first choice strategy for gene isolation. But it often is, largely because of two significant limitations of PCR:

(1) To design a PCR experiment you need some information on the sequence of the gene that you wish to isolate. This is because the PCR is primed by the pair of oligonucleotides that anneal either side of the DNA segment to be amplified. The sequences of these oligonucleotides must be complementary to their annealing sites, which means that the sequences of at least the regions bordering the gene of interest must be known. This requirement prevents the use of PCR for isolation of a gene that has never been studied before.

11

(2) It is difficult to amplify long stretches of DNA by PCR, because the polymerase cannot indefinitely extend an oligonucleotide primer. PCRs up to 3 kb long are relatively easy to perform and fragments up to 10 kb can be amplified, with practice, by standard techniques. Special procedures are needed for longer sequences, with 40 kb being the approximate limitation for even the most skilled operator.

These considerations mean that the choice between cloning and PCR can be stated as follows. If the DNA segment that you wish to isolate is less than 3 kb long and you have sufficient sequence information to be able to design suitable primers, then use PCR. If the DNA segment is longer than 3 kb and/or you cannot design primers, then use cloning.

4.4 Basic techniques needed for cloning and PCR

Whether you decide on an approach based on cloning or on PCR, you will inevitably use the same set of basic techniques—the ones covered in Volume I of this book—in your molecular biology project. The following sections provide a brief synopsis of these techniques and attempt to describe the context within which each one is used.

4.4.1 Handling bacteria (Chapter 2)

One of the most important practical skills required by the molecular biologist is the ability to grow pure cultures of bacteria, including cultures infected with bacteriophages. This is because most gene cloning experiments use the bacterium *E. coli* as the host organism. Even if the long-term objective is to clone a gene in an organism other than *E. coli* the initial manipulations will be carried out in *E. coli* because of the ease with which this bacterium can be handled. It is also quite difficult to design a molecular biology project that is based entirely on PCR, so even if PCR is your chosen approach to gene isolation you will almost certainly carry out one or more cloning experiments at a later stage of the project.

The key aspect of practical microbiology is sterile technique, which is the means by which cultures are kept pure and contamination of work surfaces and personnel is avoided. Even when carrying out manipulations that do not involve bacteria it is advisable to use sterile materials (e.g. pipette tips, glassware and microfuge tubes) and a clean technique (wear gloves, observe standard biological lab practice, etc.) as bacterial contamination and secretions from the skin can degrade DNA and RNA and inactivate restriction (and other) enzymes.

4.4.2 Preparation of DNA (Chapter 3)

For many, the first real molecular biology to be attempted is purification of DNA from the organism being studied. This provides the material to be fragmented and inserted into the cloning vector or to act as the template for a PCR. The main consideration is to obtain DNA that has not been broken up too extensively during the purification procedure and which is free of contaminants that will interfere with the subsequent enzymatic reactions, either the PCR itself or those enzymes involved in constructing recombinant DNA molecules prior to cloning. In general terms, DNA purification procedures are similar for all source organisms, although appropriate steps have to be taken to obtain an adequate cell extract and to remove contaminants such as lipids and carbohydrate.

DNA preparations are also needed to purify recombinant vector molecules in order to make further studies of the cloned gene. The vector DNA itself is usually purchased from commercial suppliers (see p. 233–236) although it is quite easy (and much cheaper) to maintain a stock of bacteria containing the vector and to prepare DNA as and when needed.

4.4.3 Preparation of RNA (Chapter 4)

The preparation of pure, intact RNA is relatively difficult because of the ubiquity of contaminating enzymes that degrade RNA molecules. RNases present in the cells from which the RNA is being extracted can be inactivated during the purification procedure by a suitable RNase inhibitor, but considerable problems can be presented by glassware and solutions contaminated with RNases derived from skin secretions. These enzymes are very resistant to physical treatments—some even withstanding autoclaving—and their control can be a problem. However, for some types of project RNA preparation is of equal or greater importance than DNA preparation. Often, the most suitable approach to obtaining a clone of your gene is to start not with DNA but with RNA. This is because the mRNA molecules present in a particular tissue or population of cells represent only those genes that are actually being expressed, which may of course be a very small subset of the total gene content of the organism. Purification of the mRNA fraction followed by conversion to cDNA (Volume II, Chapter 3) provides fragments suitable for insertion into a cloning vector, and may result in a clone library from which identification of the desired gene is much easier than from a library representing the whole genome because the number of clones that must be screened is smaller. RNA can also be used instead of DNA as the starting material for PCR (Volume II, Chapter 7). If you intend taking one of these routes in your project then it is advisable to purchase some yeast RNA (or similar) and establish that your RNase-free technique is sufficiently sound before starting work with your own, more precious RNA samples.

4.4.4 Separating DNA and RNA by gel electrophoresis (Chapter 5)

Separating DNA and RNA molecules of different sizes is achieved by electrophoresis in an agarose or, more rarely, polyacrylamide gel. The theoretical basis of gel electrophoresis is complicated but has to be tackled in order to understand the influences that factors such as gel type, buffer, size of molecule and the electrical parameters have on the separation of DNA and RNA. Many different types of agarose are now commercially available and a knowledge of these physical features of gel electrophoresis must be appreciated in order to choose the appropriate type for a particular application.

Agarose gels are routinely run to estimate the sizes of DNA and RNA molecules, which enables the success of a preparative technique or enzymatic manipulation to be assessed. Agarose gel electrophoresis is also the preliminary to analysis of DNA molecules by techniques such as Southern hybridization (Volume II, Chapter 5). These applications will be continually referred to throughout the two volumes of this book. Polyacrylamide gels, which are used to resolve small DNA molecules in procedures such as manual DNA sequencing (Volume II, Chapter 6) and transcript mapping (Volume II, Chapter 9), are more difficult to set up than agarose gels, but because of their use in protein biochemistry are often familiar to the biochemist starting to work on DNA.

4.4.5 Purifying DNA molecules from electrophoresis gels (Chapter 6)

In some areas of molecular biology there are a number of alternative procedures for achieving the same objective. The purification of DNA molecules from gels is one such case. Frequently, an agarose or polyacrylamide gel will provide a band that can be confidently identified as containing a DNA fragment of interest. The next step might be to excise the band and purify the DNA molecules prior to construction of recombinant DNA molecules and further study. This is not as easy as it might sound because electrophoresis buffers and the agarose itself contain ions and other chemicals that can co-purify with the DNA fragment and prevent, or at least retard, subsequent enzymatic manipulations. Various methods have been devised to get around the problems and the individual strengths and limitations should be understood before choosing one for a particular operation.

4.4.6 Construction of recombinant DNA molecules (Chapter 7)

So far we have dealt only with procedures involved in preparation and handling of DNA and RNA molecules. The first real step in a gene cloning experiment (see *Figure 1*) is construction of recombinant molecules by insertion of DNA fragments into vector molecules. At this stage one of the characteristic features of molecular biology research, the use of purified enzymes to carry out controlled manipulations of DNA molecules, is first encountered,. The two main types of enzyme used in recombinant DNA construction are restriction endonucleases and ligases. Restriction endonucleases (see *Appendix 3*) make double-stranded breaks at specific recognition sequences within DNA molecules. They enable DNA molecules to be cut ('restricted' in molecular biology jargon) in a reproducible fashion to provide a family of predictable fragment sizes. A circular DNA molecule that possesses just one restriction site for a particular endonuclease will be converted into a linear molecule by treatment with that enzyme. As can be seen from *Figure 1*, this is the essential manipulation required to prepare the cloning vector for insertion of the DNA fragments to be cloned. Restriction is also frequently used to analyse the results of a PCR. For example, when it is known that the amplified fragment contains a restriction site at a particular position, restriction can be used to check that the correct fragment has been amplified.

DNA ligases (*Appendix 4*) can join together DNA fragments prepared by restriction. To be ligated a pair of ends must be compatible, which means they must both be flush-ended (also called 'blunt-ended'), or both must have single-stranded overhangs ('sticky ends') that are able to base pair. The DNA fragments do not have to be prepared with the same restriction enzyme in order for their ends to be compatible: any two enzymes that produce blunt ends will yield fragments that can be ligated, and two enzymes with different recognition sequences may still produce compatible single-strand overhangs. In addition, one type of sticky end can be converted to another type of sticky end or into a blunt end, and blunt ends given sticky overhangs, by use of the more specialized manipulative techniques described in Chapter 7.

4.4.7 Introduction of recombinant molecules into host cells and recombinant selection (Chapter 8)

Once recombinant DNA molecules have been constructed they must be introduced into *E. coli* cells. This is quite straightforward and standard techniques exist for different types of vector. The means used to identify colonies containing recombinant molecules are more complex. Only a proportion of the cells in the culture will take up any DNA molecule at all, and of these a fraction will usually take up a self-ligated vector that carries no insert DNA. Identification of recombinant colonies (i.e. those containing a recombinant DNA molecule) makes use of the genes carried by the cloning vector. One of these will have been inactivated by insertion of the cloned DNA, leading to a phenotypic change that can be directly selected (e.g. antibiotic resistance/sensitivity) or easily scored (e.g. ability to use a chromogenic substrate).

Under some favourable circumstances it may be possible at this stage to identify recombinant clones of interest, This would be the case if the recombinant molecules were constructed in such a way that each one must contain the correct gene, or if the cloned gene itself confers an identifiable phenotype on the host cell. However, this is not usually the case and more frequently the cloning experiment will simply provide a collection of recombinants from which the desired one must be identified by a second series of experiments, using the strategies described in Volume II, Chapter 1.

4.4.8 Cloning vectors (Chapter 9)

By now it will be clear that the choice of cloning vector is an important one. It will determine the nature of the procedure used to introduce the recombinant molecules into the host cells and it will specify the way in which recombinants are selected. It is important to choose a vector that is suitable for the size of DNA fragments that you wish to clone, as some vectors are designed for relatively small fragments (5 kb and less) whereas others require fragments above a certain size.

There are many different cloning vectors for use with *E. coli* but all are derived from natural bacterial plasmids or from the genomes of bacteriophages λ or M13. Plasmids and phage genomes possess replication origins and either code for replicative enzymes or can use host enzymes to direct their own replication.

They therefore possess the means for existing and multiplying within their host cells, providing the basic features needed for obtaining a recombinant clone.

Choice of cloning vector is difficult and advice from different sources can be contradictory. Over 100 different types are available commercially, and a whole host of applications are catered for. Often, it is tempting to use a sophisticated vector even though it carries a number of functions that are unnecessary for the experiment being conducted. As a rule of thumb, choose the simplest vector suitable for your requirements, but be prepared to change if you encounter problems with handling or recombinant selection. Once you have success with a vector try to use it for future experiments whenever possible: better the vector you know than the vector you don't.

5 Planning an informative project

To achieve success in molecular biology it is clearly important to understand the techniques and attain the necessary practical skills. It is also important to understand what can be achieved by a particular research programme. Molecular biology is not a panacea; it can be very difficult to interpret in a meaningful way the results of a molecular biology project if there is no complementary bio-chemical or genetic information on the system being studied.

To illustrate the potential pitfalls of molecular biology projects we will take a fairly typical scenario. Consider a research group that has studied an enzyme for a number of years and now wishes to clone the gene, sequence it, and study its expression. Before starting this new project several questions should be asked. Similar questions are appropriate before projects with other starting points or objectives. The points made here will be expanded on in Volume II, Chapter 1, but it is important that the questions are asked right at the beginning of the project.

5.1 Will it be possible to identify the correct clone?

Only under special circumstances will it be possible to devise the cloning experiment in such a way that the desired gene can be identified immediately. Normally, the primary objective will be to prepare a clone library, either from cDNA or from genomic DNA, and to identify the clone containing the relevant gene from this library. To do this, a probe specific for the desired gene will be needed. Is such a probe available?

In general terms there are four ways of obtaining a suitable probe:

(1) The amino acid sequence of the enzyme, if known, can be used to dictate the sequence of a specific oligonucleotide hybridization probe.

(2) If the gene for the enzyme has already been cloned from a related organism, then this gene can be used as a heterologous hybridization probe.

(3) If the enzyme is known to be abundant in a particular tissue, then it may be possible to identify the relevant clone by virtue of its abundance in a tissue-specific cDNA library.

(4) If the enzyme has been purified, then an antibody raised against it can be used to screen recombinant cells for the presence of the enzyme.

These are generalizations, but they cover most of the practical ways of identifying a gene in a clone library. If one of the above possibilities cannot be satisfied then it would be pointless to embark on a gene cloning project.

5.2 Will the DNA sequence provide any new information?

A DNA sequence on its own is not much use. If you do not know what to expect then it may not even be possible to assign unambiguously the coding region of a gene. Intron boundaries can be guessed from their similarities to consensus sequences but often two or more alternative positions for an intron cannot be discounted purely from sequence information. Therefore, it may not even be possible to determine the coding sequence of the gene.

If the amino acid sequence of the enzyme is already known, then the codon sequence will be identifiable, and any introns can be located with precision. However, if the amino acid sequence is already known then what new information will you get from the gene sequence?

What if the gene sequence can be identified but the function of the translation product is unknown. It may be possible to identify the function of a gene sequence by comparing it with other sequences held in the DNA databases, but often a sufficiently convincing match will not be found, although this problem is becoming less severe with the gradual accumulation of complete genome sequences for different types of organism. Nevertheless, how to go from DNA sequence to identification of a gene product is still one of the great problems of molecular biology research and can provide an effective block to the progress of a project. This should be borne in mind if cloning and sequencing are to be used to study, for example, tissue-specific genes for which translation products have not been identified, or genetic loci whose phenotypic effects are not clearly associated with a known gene product.

So far we have not made too much progress with our project. What about expression of the gene?

5.3 Will studies of gene expression be informative?

Before you start to purify DNA you can assume that your gene has a promoter and, if it is differentially expressed, other upstream control sequences to mediate gene regulation. Will your expression studies take you any further than this? The answer is yes, probably, although as with DNA sequencing a biologically meaningful interpretation of the results depends on being able to relate the data provided by molecular biology to information from biochemistry or genetics. Without being too adventurous it is possible to:

• Determine in which tissues a gene is expressed

• Determine the approximate sizes of transcripts and whether the transcripts are the same size in all tissues

- Determine the precise 5′-and 3′-termini of the transcripts and of any introns within the transcripts

More ambitious experiments can:

- Locate upstream control sequences that mediate gene regulation
- Isolate regulatory proteins that bind to these control sequences

However, if results along these lines are achieved in a vacuum then their importance will relate primarily to molecular biology and only vaguely to the biology of the cell as a whole. Can gene regulation be related to control of activity of your enzyme, or are transcriptional events masked by allosteric controls and protein activation events? Are you aware of external and internal biochemical stimuli that influence gene expression? Are there clearly-defined defects in gene product activity that may point to impairments in gene regulation? If you can ask specific questions along these lines then you will be able to make the difficult transition from genes to the rest of the cell. If you have no interesting questions then you should ask yourself why you wish to embark on a molecular biology project.

5.4 Conclusion

In the preceding sections I have deliberately taken the stance of devil's advocate, not to dissuade you from a molecular biology project, but to try to ensure that at the outset the foundations are laid for a successful research programme. Molecular biology can provide a wealth of information and the chapters in this book will help you obtain that information. It is up to you to make sure the information is useful.

References

1. Brown, T. A. (ed.) (1998). *Molecular biology labfax*, 2nd edn, Vol. 2. Academic Press, London.
2. American Chemical Society, Committee on Chemical Safety (1985). *Safety in academic chemistry laboratories*. American Chemical Society, Washington.
3. Collins, C. H. (ed.) (1988). *Safety in clinical and biomedical laboratories*. Chapman and Hall, London.
4. Collins, C. H. (ed.) (1999). *Laboratory-acquired infections: history, incidence, causes and prevention*, 4th edn. Butterworth-Heinemann, Oxford.
5. Collins, C. H. (1995). *Collins and Lyne's microbiological methods*, 7th edn. Butterworth-Heinemann, Oxford.
6. Health and Safety Commission (1993). *Control of substances hazardous to health; and control of carcinogenic substances*, 4th edn. HMSO, London.
7. Health Services Advisory Committee (1991). *Safe working and the prevention of infection in clinical laboratories*. HMSO, London.
8. Imperial College of Science and Technology (1974). *Precautions against biological hazards*. Imperial College, London.
9. Health and Safety Executive (1996). *A guide to the genetically modified organisms (contained use) regulations 1992, as amended in 1996*. HMSO, London.
10. Anonymous (1984). *Federal Register*, 49, Part VI, Number 227, 46266.

11. Health and Safety Commission (1982). *The ionising radiations regulations*. HMSO, London.

12. Health and Safety Commission (1985). *The protection of persons against ionising radiation arising from any work activity*. HMSO, London.

13. Anonymous (1993). *The Radioactive Substances Act 1993*. HMSO, London.

14. Coggle, J. E. (1983). *The biological effects of radiation*, 2nd edn. Taylor and Francis, New York.

15. Martin, A. and Harbison, S. A. (1986). *An introduction to radiation protection*, 3rd edn. Chapman and Hall, London.

16. United Nations Scientific Committee on the Effects of Atomic Radiation (1996). *Sources and effects of ionising radiation*. United Nations, New York.

17. International Commission on Radioactive Protection (1977). *Recommendations of the International Commission on Radiation Protection*. *ICRP Publications 26 and 27*. Pergamon Press, Oxford.

18. Anonymous (1970). *Radiological health handbook*. US Department of Health, Education and Welfare, Washington, DC.

19. Meisenhelder, J. and Semba, K. (1996). In *Current protocols in molecular biology*. (ed. F. M. Ausubel, R. Brent, R. E. Kingston *et al.*), Unit A.1F. John Wiley, New York.

20. Anonymous (2000). *Life sciences catalogue*. Amersham International, Amersham.

21. Anonymous (2000). *Sigma–Aldrich material safety data sheets on CD-ROM*. Sigma–Aldrich Corp., Milwaukee.

22. Anonymous (2000). *BDH hazard data sheets, computer package*. BDH, Poole.

23. Lenga, R. E. (ed.) (1988). *The Sigma–Aldrich library of chemical safety data*, 2nd edn. Sigma–Aldrich Corp., Milwaukee.

24. Lenga, R. E. and Votoupal, K. L. (ed.) (1993). *The Sigma–Aldrich library of regulatory and safety data*. Sigma–Aldrich Corp., Milwaukee.

25. Lewis, R. J. (1988). *Sax's dangerous properties of industrial materials*, 8th edn. Van Nostrand Reinhold, New York.

26. Bretherick, L. (ed.) (1986) *Hazards in the chemical laboratory*, 4th edn. Royal Society of Chemistry, London.

27. National Research Council (1981). *Prudent practices for handling hazardous chemicals in laboratories*. National Academy Press, Washington, DC.

28. Royal Society of Chemistry (1992). *Hazards in the chemical laboratory*, 5th edn. Royal Society of Chemistry, Cambridge.

29. Furr, A. K. (ed.) (1990). *CRC handbook of laboratory safety*, 3rd edn. CRC Press, Boca Raton.

30. Lefevre, M. J. (1980). *First aid manual for chemical accidents*. Dowden, Hutchinson and Ross, Stroudsburg.

31. Pitt, M. J. and Pitt, E. (1985). *Handbook of laboratory waste disposal*. Wiley, New York.

32. Young, J. A. (ed.) (1987). *Improving safety in the chemical laboratory*. Wiley, New York.

33. Great Britain, Department of Education and Science (1976). *Safety in science laboratories*, 2nd edn. HMSO, London.

34. Pipitone, D. A. (1984). *Safe storage of laboratory chemicals*. Wiley, New York.

35. Collings, A. J. and Luxon, S. G. (ed.) (1982). *Safe use of solvents*. Academic Press, Orlando.

36. Anonymous (1975). *Toxic and hazardous industrial chemicals safety manual*. International Technical Information Institute, Tokyo.

37. Perbal, S. (1988). *A practical guide to molecular cloning*, 2nd edn. Wiley, New York.

38. Bensaude, O. (1988). *Trends Genet.* **4**, 89.

39. Lunn, G. and Sansone, E. B. (1987). *Anal. Biochem.* **162**, 453.

40. Quillardet, P. and Hofnung, M. (1988). *Trends Genet.* **4**, 89.

41. Brown, T. A. (1995). *Gene cloning: an introduction*, 3rd edn. Chapman and Hall, London.

Chapter 2

Microbiological techniques for molecular biology: bacteria and phages

Brian W. Bainbridge

Division of Life Sciences, Kings College London, 150 Stamford Street, London SE1 8WA, UK

1 Introduction: techniques for handling microbes

1.1 Basic microbiological techniques

Recombinant DNA technology depends on the manipulation of particular strains of bacteria and bacteriophages (phages). Expertise in isolating, checking, growing, and analysing these strains is crucial for the success of genetic engineering. Many experiments are now permitted under conditions of 'good microbiological technique' and anyone embarking on research in this area should check exactly what is meant by this phrase. Essentially, all of the basic techniques depend on the culturing of a particular microbe in the absence of other organisms (sterile or aseptic technique), the isolation of a genetically pure culture or clone derived from a single cell (single-colony or single-plaque isolation) and the characterization of the known genetic features of the strain. For safety, experiments should be carried out in such a way that recombinant microbes do not escape into the environment or infect the experimenter or others. The basis of sterile technique is the efficient sterilization of all equipment and growth media followed by the protection of the media from contamination during manipulations. If contamination of a culture is suspected then colony morphology, smell of the culture or staining of the cells followed by microscopic examination can be useful. Growth of the strains in liquid or solid media allows larger numbers of cells to be cultured and it is important to understand the various phases of the growth curve, as different types of cells are required for different experiments (1). Plating techniques such as dilution and streak plates can give rise to single colonies or plaques derived from individual cell or phage particles. Selective techniques are often used to isolate particular strains or to detect rare events, for example in transformation (see Chapter 8), and such techniques are very sensitive to low levels of contamination by unrelated strains which happen to have the same characteristics as the desired strain: this may be resistance to antibiotics or nutritional independence.

1.2 Basic techniques of microbial genetics

Genetic purity of strains, plasmids or bacteriophages is critical and it should be clearly understood that no strains can ever be completely stable (2). Growth can lead to rearrangements of DNA, mutation of essential genes or infection by foreign DNA. Strains should therefore be carefully checked to make sure that they are of the correct genotype before starting crucial experiments. Chemicals and enzymes are not self-replicating and they are therefore usually purchased from suppliers. In contrast, strains of bacteria and phages are frequently sub-cultured many times and passed from one research group to another. Quality control and frequent checking is necessary or significant alterations in the genome of the organism may occur. It is particularly important to make a careful note of any deliberate changes or improvements in particular strains. This means recording the pedigree of the strains and using a unique numbering system, and detailing the full genotype. The dangers here are that essential gene mutations such as *recA* (deficient for the major recombination system) may have reverted to wild type and transposons may have caused rearrangements. An empirical approach can be adopted such that if the technique works the strain must be correct, but this can delay the problem until a later stage. Unless you are absolutely sure of the origin of the strain it is much safer to obtain the correct culture from a reputable supplier. When the correct strain has been successfully characterized it is essential that it is preserved efficiently. Any growth can lead to mutation and any desiccation or contamination can lead to death of the culture. The basis of preservation is to stop growth by lyophilization or freezing in the presence of cryoprotectants such as glycerol. Duplicate cultures should be kept in different locations to avoid dangers of apparatus failure.

1.3 Safety in the molecular biology laboratory

Although *Escherichia coli* K12 is not considered to be pathogenic, there are other strains of *E. coli* which are dangerous and can cause septicaemia or kidney infections. The development of disease symptoms depends on the type and number of organisms which gain access to the body, as well as on the efficiency of the immune system of the laboratory worker. Most healthy individuals can cope with small numbers of microbes but depression of the immune system by illness or infection or the use of immunosuppressive drugs can markedly increase the risks of a serious infection. It is safer to assume that all microbes are potential pathogens and to treat them with respect. The basic safety precautions are shown below and should be followed along with local and national regulations (3, 4).

(1) Laboratory overalls, which should ideally be side-or back-fastening, must be worn.

(2) Avoid all hand-to-mouth operations, such as licking pencils or labels.

(3) Make a habit of washing hands before leaving the laboratory.

(4) Before starting work, protect all cuts with adequate waterproof dressings.

(5) Cultures spilt on the bench, floor, apparatus or on yourself or others, should be treated with disinfectant. Material used to wipe up should be discarded for incineration or sterilization.

(6) Do not eat, drink, smoke or apply cosmetics in the laboratory.

(7) All contaminated apparatus should be sterilized before washing or disposal.

(8) All contaminated glassware such as pipettes and tubes should be discarded into disinfectant prior to sterilization and washing.

(9) All contaminated disposable plasticware should be discarded into strong autoclavable bags for sterilization or incineration.

(10) All apparatus for autoclaving or incineration should be carried in leak-proof containers.

(11) All apparatus used for the culture of microbes should be clearly labelled before inoculation. Apparatus to be left in communal areas should show your name, the organism and the date.

(12) All contaminated syringe needles should be discarded into special 'sharps' containers for special sterilization and disposal.

(13) Every effort should be made to avoid the production of aerosols. These are produced, for example, by blenders, centrifugation, ultrasonication and movement of liquids against surfaces. If the microbes present a hazard then the equipment used should be placed in a suitable safety cabinet.

(14) Records should be kept of the storage and transport of all microbes.

A more detailed treatment of this subject will be found in reference works by Collins (5, 6). In the UK, the regulations for the Control of Substances Hazardous to Health (7) clearly define microorganisms as a health hazard and state that an assessment should be made of the risks created by microbiological work and of the steps that need to be taken to ensure safe practice. Essentially, exposure to microbes by inhalation, ingestion, absorption through the skin or mucous membranes or contact with the skin must be either prevented or controlled. Thus, safety should be treated seriously and precautions taken to limit the access of microbes to the environment, laboratory workers, and the public. Safety is not the blind following of regulations but an awareness of the hazards and the methods which can be used to minimize them.

This chapter is intended to provide a basic introduction to the techniques involved in the handling of *E. coli* and its phages λ and M13. It does not include techniques in microbial genetics such as mutant induction, gene mapping and replica plating, which will be found in a variety of other manuals (8, 9).

1.4 Sterilization and disinfection

It is important to distinguish between these two processes. The aim of sterilization is to eliminate all microbes from laboratory equipment or materials, whereas disinfection aims to eliminate organisms which may cause infection. Space does not permit a theoretical treatment of this subject (5) but the important practical principles will be given.

Sterilization can be achieved by heat, chemicals, radiation, or by filtration. Nichrome loops are sterilized by flaming in a Bunsen burner, and disposable plastics can be sterilized by incineration. Glassware can be sterilized either by autoclaving or by dry heat. Autoclaving which uses wet heat is much more efficient than dry heat as hydrated microbes are killed more easily. Autoclaves vary from domestic pressure cookers to large, industrial-size motorclaves. It is very important that the operating instructions are followed and in particular that the autoclave is not overloaded, as the central region may not reach the necessary temperature. It is important to remove all of the air from the auto-clave because the presence of air will depress the final temperature reached. Autoclave tape which changes colour after the correct time and temperature is a useful check. Loosen the caps of all bottles and do not autoclave completely-sealed bags in small autoclaves. Always make sure that there is sufficient water in pressure cookers and check that the correct procedure for autoclaving and recovery of materials is followed. Autoclavable plastic tubes such as pipette tips and microfuge tubes should be wrapped in autoclavable nylon bags. Dry-heat sterilization is normally used for flasks and glass pipettes, which should be left on a 6–12 h cycle at 160 °C. Sterilization with chemicals and radiation are impractical in the average laboratory. Sterile plasticware is normally produced by γ irradiation.

Disinfection procedures will vary from laboratory to laboratory: 2% (v/v) Hycolin (Adams Healthcare) is a general-purpose disinfectant that turns from green to blue when no longer effective; an alternative is Virkon (Philip Harris). Any contaminated glassware or unwanted cultures should be immersed in disinfectant before autoclaving and washing. Contaminated disposable plastic-ware should be placed in autoclavable bags or bowls, which must be leak-proof. This material should be disposed of by incineration. Detailed procedures are the responsibility of the local safety officer.

1.5 Basic principles of aseptic technique

There are two basic principles of aseptic technique: protection of yourself and others, and protection of cultures and apparatus from contamination by un-wanted microbes. In normal laboratory areas, microbes are everywhere—in the air, in dust, on your fingers. It is very difficult to produce an environment that is completely free from microbes, and special equipment such as laminar-flow cabinets and sterile areas is required. In a clean laboratory, with reasonable precautions, it is not necessary to use inoculating cabinets for the preparation of media and the manipulation of cultures. However, if you have trouble with contamination or your cultures are particularly slow-growing then a cabinet can be useful. Safety cabinets should not be needed for experiments classified as needing only 'good microbiological technique'. Experiments at higher levels of containment need different precautions in different countries (3, 4) and you should consult your local safety officer for guidance.

Unless laboratory air has been efficiently filtered, it will contain many suspended bacterial cells and spores, fungal spores, and, in some laboratories, air-stable phage particles. This population of particles is added to by air movements which resuspend dust particles from bench surfaces. These airborne particles will settle on to any exposed surface and this is a major source of contamination, therefore anything which is to be kept sterile must be exposed to the air for a minimum period of time. Dust and aerosol particles tend to settle rather than drift sideways unless there is a draught of air. Consequently, containers and Petri dishes should not be left open, surface upwards, although tubes opened at an angle are less at risk. All apparatus which cannot be flamed in a Bunsen burner immediately prior to use should be left in the wrappers or containers in which they have been sterilized until needed. No sterile equipment should be allowed to come into contact with non-sterile surfaces. Plugs and caps from sterile tubes and bottles should not be placed on the bench, although they can be placed on a tile swabbed with a disinfectant.

The commonest source of contamination in the laboratory is the access of non-sterile air to the apparatus. This is increased by draughts and general movement of air and it follows that every effort should be made to work in still-air conditions. Windows and doors should be closed and all rapid movements in the laboratory should be eliminated. It is obvious that the laboratory should be free from dust which could be resuspended by air movement but it should be remembered that cleaning techniques such as sweeping and dusting can be a serious source of aerial contamination. The major advantage of inoculating cabinets is that they give protection from these air movements and allow a small volume of air to be sterilized by UV radiation or by a 70% (v/v) alcohol spray. The principles of spraying the air in a cabinet or over a bench is that it settles dust particles on to the bench, from where they can be removed with a paper towel. Fungal spores are adapted to aerial transmission and particular care should be taken when handling fungi or disposing of apparatus contaminated by fungi. Fungal spores released into the environment can take up to 7h to settle and therefore can be a source of contamination for a considerable period.

Skin, hair, breath, and clothing are all sources of microbes and it is particularly important that you do not touch sterile surfaces such as the tips of pipettes and the inside of containers. Do not bend over your equipment such that skin scales or dust from your hair might fall into your cultures. Problems have arisen from contamination with yeasts traced to home baking of bread. Where strict asepsis is required (as in operating theatres) sterile caps, gloves and gowns should be worn.

2 Culturing of *Escherichia coli*

A culture of an *E. coli* strain will normally be received as a broth culture, on a Petri dish or as a freeze-dried culture from a supplier. The first step is to make a careful record of the strain number and genotype of the strain. From this you will be able to identify an appropriate medium on which it will grow well and any additions such as antibiotics which are necessary to ensure the stability and maintenance of plasmids. Prepare some well-dried agar plates, as described in *Protocol 1*. It is necessary to dry the plates because *E. coli* is motile and will swim across the plate in the thin film of water. In addition, contaminants will also spread more easily across the plate and the desired single colonies will not be isolated. Cooling the agar at 50 °C reduces the condensation of water on the inside of the lids. Drying the plates overnight at 37 °C has the advantage that contaminated plates can be detected and discarded, but care should be taken to avoid microcolonies on the plate. This is particularly critical with spread plates where the contaminant colonies will be spread over the plate. To avoid this problem it is better to dry the plates in an oven.

Protocol 1

Preparation of agar plates

Equipment and reagents

- Agar medium
- Steamer or microwave oven
- Water bath at 50 °C
- Petri dishes

Method

1 Select the ingredients for the required agar medium.

2 Loosen the top of the agar bottle and melt the agar in a steamer or in a microwave oven. Do not use metal caps in a microwave oven.

3 Swirl the liquid gently to check that the agar is fully melted. Take care that super-heated agar does not boil over.

4 Allow the agar to cool for about 10 min at room temperature and then place in a water bath at 50 °C for at least 20 min.

5 Place the concentrated nutrient medium in the same water bath to equilibrate to the same temperature.

6 Flame the tops of the glass bottles containing agar and concentrated nutrients and pour the nutrient liquid into the bottle containing the agar. Screw the cap tight and shake to ensure complete mixing.

7 Return the bottle to the water bath and allow time for any air bubbles to disappear.

8 Arrange the sterile Petri dishes on a level surface and label the base of each plate to indicate the medium prepared.

9 Remove the bottle from the water bath and wipe the outside carefully with a paper towel (water baths can be contaminated).

10 Flame the neck of the bottle and pour the required amount of medium into the plate. This will vary between 10 ml for a thin plate for short-term bacterial culture to 40 ml for phage culture where large plaques are required.

Protocol 1 continued

11 Allow the plates to set. Dry the surface of the plates by overnight incubation at 37°C (check for contamination the following morning) or by opening the plates and placing them, medium surface down, in an oven at 45–55°C. The lids should also face downwards separately from the base of the plate. Leave plates at 45–55°C for 15 min.

12 Most nutrient plates can be stored for at least a few weeks at 4°C following wrapping in parafilm or sealing in a plastic bag.

Escherichia coli can be cultured on slopes, on plates, in broth, or in stab cultures. The first step is to isolate single colonies usually by a streak plate method (see *Protocol 2*). A single colony is used to produce a series of identical broth cultures and a check is made of the phenotype of the strain. The culture can then be used for experimental purposes. As soon as it is determined that the culture is correct, every effort should be made to preserve it. This can be done on plates, in broths, or in stab cultures where protection from contamination and desiccation is particularly important (Section 4). Freezing is necessary for long-term storage.

2.1 Single-colony isolation

The principle of this technique is to streak a suspension of bacteria until single cells are separated on the plate. Each individual cell will then grow in isolation to produce a clone of identical cells known as a colony. The vast majority of these cells will be genetically identical, although mutation can occur during the growth of even a single colony to give low levels of mutant cells. This technique assumes that there are no clumps of cells in the culture; however, it is not unknown for contaminant organisms to stick to bacterial cells via an extracellular polysaccharide. Any clumps will be visible upon examination of a culture under the microscope (see Section 3.8). If they are present, every effort should be made to disrupt them by suspension in PBS followed by agitation or gentle ultrasonication treatment. Repeated single-colony isolations should result in a pure culture.

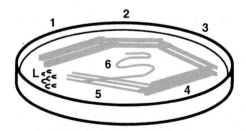

Figure 1 Procedure for the production of single colonies by the streak plate technique, as described in *Protocol 2*.

Protocol 2

Streak plate method for single-colony isolation (refer to *Figure 1*)

Equipment and reagents

- Nichrome loop
- Mixer

- *Either* PBS (8.0 g NaCl, 0.34 g KH$_2$PO$_4$, 1.21 g K$_2$HPO$_4$ per 1 l). pH should be 7.3; sterilize by autoclaving
- *or* Ringer's solution

Method

1 Flame a Nichrome loop which is about 3 mm across and has a stem of about 6 cm. Allow the loop to cool, or cool by immersion in a sterile area of agar.

2 Flame the neck of an overnight broth culture and remove a loopful of cells. Alternatively, make a suspension of cells directly from an agar plate in PBS or dilute Ringer's solution. Vortex and remove a loopful of cell suspension.

3 Streak the cells at one side of a well-dried agar plate at position 1 as shown in *Figure 1*. Streak several times close together.

4 Flame the loop and cool carefully at one side of the plate (position L).

5 Streak again at position 2 on *Figure 1*.

6 Flame the loop and cool as before. Repeat steps 4 and 5 as indicated at positions 3, 4, 5 and 6 on *Figure 1*.

7 Incubate the plate at 37 °C with the agar facing downwards to minimize contamination and to reduce the chance of droplets of condensation falling on the agar surface.

The plates are then examined for colony morphology and the presence of possible contaminants. If all of the colonies are of a uniform size and appearance it can be assumed that you have a pure culture. Subculture of almost any colony should give the required strain and the plate is worth keeping as a future source of a purified culture. When plasmid-containing strains are plated on a medium supplemented with an antibiotic it is often observed that colonies of a variety of sizes are obtained. It is not always immediately obvious which is the correct strain and it may be necessary to subculture a representative range of colonies on to fresh medium for further checking. A careful note should be made of the characteristics of the correct strain for future reference. Re-streak the correct colony to give a plate containing a uniform colony size and use this as a source of purified culture.

In a number of experiments, such as transformation and the preparation of plasmid or cosmid genomic libraries, it is necessary to have as many single colonies on the same plate as possible. This involves a dilution series and spread plates:

Protocol 3

Dilution series and the spread-plate method for single-colony isolation

Equipment and reagents

- PBS or Ringer's solution (see *Protocol 2*)
- LB agar plates (10 g bacto-tryptone, 5 g bacto-yeast extract, 10 g NaCl, 15 g bacto-agar per 1 l). Check the pH and adjust to

7.0–7.2 with NaOH; sterilize by autoclaving at 121 °C, 103.5 kPa (15 lb/in²), for 20 min
- Glass spreader and ethanol

Method

1   Prepare a series of six tubes containing 9 ml of PBS or dilute Ringer's solution. Label the tubes –1, –2, –3, –4, –5, and –6. Prepare three LB agar plates labelled –4, –5, and –6.

2   Take 1 ml of the test culture (assumed to be about 10^8 cells/ml) and add to tube –1. This is a 10^{-1} dilution. Vortex the tube to mix the cells.

3   With a fresh pipette or disposable tip, take 1 ml from the –1 tube and add to the –2 tube. This is a 10^{-2} dilution.

4   Repeat this operation from –2 to –3 and so on down the series until tube –6 is reached.

5   Take 100 µl of the –6 dilution and add to an LB agar plate.

6   Dip a glass spreader into ethanol in a glass Petri dish and pass rapidly through a Bunsen burner to burn off the alcohol.

7   Cool the spreader on the agar surface and spread the suspension evenly over the plate. Repeat for tubes –5 and –4.

8   Incubate the plate overnight at 37 °C. There should be about 1000, 100, and 10 colonies on the –4, –5, and –6 plates respectively. Plate lower dilutions for more dilute suspensions of cells.

9   A variation of this method, which is more economical on plates, is to add drops of 20 µl on marked places on the same plate without spreading. This is the Miles–Misra technique. The plate should be incubated at 25 °C for 2 days to give smaller colonies, which are easier to count and to subculture.

2.2 Small-scale broth culture

Suspend an appropriate single colony in 0.5 ml of PBS or dilute Ringer's solution (see *Protocol 2*) and vortex the suspension. Use this to inoculate a series of three or four tubes containing 5 ml of a suitable liquid nutrient medium. Incubate the tubes at 37 °C overnight and then store the tubes at 4 °C until required. If higher cell densities are required the tubes can be shaken at 250 r.p.m. during overnight growth. These cultures can be used over a period of weeks to provide a source of purified cultures. Alternatively, before each experiment take a single colony from the original plate and use it to inoculate a single 5 ml broth culture, shake overnight and then use to seed a larger-scale culture.

2.3 Large-scale broth culture

These cultures are prepared when more cells are required for plasmid extraction or as a source of competent cells for transformation. The scale of the culture depends on the number of cells required. A typical yield would be 5×10^8 cells/ml so that a 20 ml culture would give a total of 10^{10} cells (dry weight 4 mg) whereas a 500 ml culture would produce 2.5×10^{11} cells (dry weight 100 mg). A 3–5 ml culture is suitable for a mini-preparation of plasmid DNA whereas a 500 ml culture is required for a large-scale plasmid preparation (see Chapter 3, Section 2.3).

Protocol 4

Large-scale broth culture of *E. coli*

Equipment and reagents

- Sterile media
- Spectrophotometer and cuvettes

Method

1 Prepare a 5 ml broth culture from a single colony as described above (Section 2.2) and use this as a seed culture for the large-scale culture.

2 Prepare conical flasks containing the appropriate volume of sterile medium in flasks roughly 10 times the volume of the medium (i.e. 25 ml in a 250-ml flask or 200 ml in a 2-l flask). This ratio has been shown to give maximum aeration so that oxygen is not the growth-limiting factor. However, it is possible to obtain good yields of cells using 500 ml of broth in a 2-l conical flask.

3 Dilute the seed culture 1 in 20 into the large-scale culture and shake the flask overnight at 37°C on a rotary shaker at 250–300 r.p.m.

4 Harvest the cells by centrifugation at 2000–4000 g, depending on the tightness of the pellet required and the characteristics of the strain.

5 Resuspend the pellet of cells in a suitable buffer—this will vary according to the purpose of the experiment.

6 Recentrifuge and resuspend the pellet for further use.

7 To follow a growth curve (see Section 2.3.1), wait until the culture is just visibly turbid (usually about 2–3 h) and remove a small sample (1–2 ml depending on culture size).

8 Assay OD at 550 nm in disposable cuvettes using the uninoculated growth medium as a blank.

9 Continue sampling approximately every 30 min until the desired OD is reached.

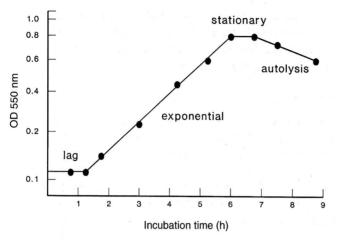

Figure 2 Typical growth curve of *Escherichia coli* in a shake flask at 37°C. Samples are taken about every 30 min and the OD at 550 nm measured. The length of the lag, stationary and autolysis phases will vary.

2.3.1 The bacterial growth curve

A knowledge of bacterial growth kinetics is essential for a number of techniques in recombinant DNA technology (1). *Figure 2* shows a typical growth curve for *E. coli*, the basic features of which are a lag phase of about 1.5 h followed by a period of exponential growth, a deceleration phase, a stationary phase and finally a decline or autolytic phase. When an overnight culture is diluted 1 in 20 into fresh medium, there is typically a lag phase when no growth can be detected. This can vary from 1.5 h to 3 h, depending on the strain, its growth rate, and the number of cells inoculated. The easiest way to monitor growth is by use of spectrophotometry. Essentially, this is a measure of the turbidity (OD) of the

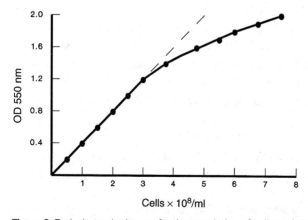

Figure 3 Typical standard curve for the correlation of cell numbers and OD. Cell numbers can be estimated by the dilution and spread-plate techniques described in *Protocol 3*. The standard curve will vary with strain, cell size and spectrophotometer used. It is useful to obtain dry weight estimates for a similar range of samples.

culture estimated on instruments which record absorbance. When a beam of light is passed through a suspension of bacterial cells, light is scattered so less light passes through the suspension in comparison with a control sample. The intensity of the light decreases exponentially as the bacterial concentration increases linearly over a limited range. Ideally, this should be checked for a particular strain with a known spectrophotometer. In a complete analysis there should also be a check of OD against cell numbers or dry weight of cells (*Figure 3*). In practice, it is normally sufficient to note values of OD for particular strains which give good competent cells or adequate yields of cells for plasmid extraction or growth of phage λ. Until you are used to a particular strain it is recommended that you follow OD at intervals and plot this on semi-log paper. This will show the lag and exponential period of growth and it will allow you to predict when the cells will be ready for harvesting.

3 Characterization of bacterial strains

3.1 Genotypes and strain nomenclature

The first step in characterizing a strain is to check its phenotype and, indirectly, its genotype. Some expertise is required in understanding symbols, which in general are based on a system proposed by Demerec *et al.* (10). A summary of the major points of this proposal follows (see also ref. 11).

(1) Each locus of a wild-type strain is designated by a three-letter, lower-case italicized symbol (e.g. *arg* is the gene determining and regulating arginine biosynthesis).

(2) Different loci, any one of which may mutate to produce the same gross phenotypic change, are distinguished from each other by adding an italicized capital letter immediately following the three-letter lower-case symbol (e.g. *argA*, *argB*).

(3) A mutation site should be designated by placing a serial isolation number after the locus symbol. If it is not known in which of several loci governing related functions the mutation has occurred, the capital letter is replaced by a hyphen (e.g. *argA1*, *argB2*, *arg-3*).

(4) Plasmids should be designated by symbols which are clearly distinguishable from symbols used for genetic loci.

(5) Mutant loci and mutational sites on plasmids should be designated by symbols of the same type as those used for loci on the chromosome.

(6) Phenotypic traits should be described in words, or by the use of abbreviations which are defined the first time they appear in a given paper. The abbreviations should be clearly distinguished from the genotype symbols (e.g. the Arg⁻ phenotype is an arginine requirement, associated with the *argA* locus).

(7) Strains should be designated by simple serial numbers. To avoid duplications, different laboratories should use different letter prefixes. Strain designations should not be italicized (e.g. HB101).

(8) When a strain is first mentioned in a publication its genotype should be described and relevant phenotypic information should be given. The genotype includes a list of all mutant loci and/or mutant sites, a list of plasmids, and information concerning the state and location of any episomes (plasmids or prophage).

A careful record should be kept of strain designations, the origin of the strains and details of the genotype. It is not enough simply to label the strain with the plasmid which it contains as you will need to grow the strain under conditions that allow good growth without the loss of the plasmid. Maintenance of the plasmid often depends on a selective technique; for example, by incorporating an antibiotic such as ampicillin in the medium (see Section 3.3). In some cases, a wild-type gene on the plasmid will complement a deletion or mutant gene on the chromosome. In these cases, growth on a medium deficient in the growth factor will be necessary to ensure survival of the plasmid (Section 3.2). Similarly, the knowledge that the host strain possesses a deletion of the lactose operon may be essential in analysing plasmids which carry the normal β-galactosidase gene. These various points are illustrated by considering the genotype and phenotype of *E. coli* strain HB101.

F⁻   Lacks the sex factor F.

hsdS20 A mutant of type ($r_B^- m_B^-$) in the site recognition gene for the strain B restriction endonuclease system. This makes the strain deficient for both restriction and modification of the DNA.

recA13 Deficient for major recombination protein. The strain is UV sensitive and lacks the major *E. coli* recombination system. This reduces the chance of rearrangements and transfer of recombinant DNA.

ara-14 Unable to utilize arabinose as the sole source of carbon and energy.

proA2 Requirement for proline in the medium.

lacY1 Mutation in the permease for the uptake of lactose.

galK2 Unable to utilize galactose.

rpsL20 Resistance to streptomycin (Sm^r) owing to a mutation in the ribosome.

xyl-5 Unable to utilize xylose.

mtl-1 Unable to utilize mannitol.

supE44 Carries a suppressor for the amber chain-terminating triplet.

(λ)⁻   Non-lysogenic for bacteriophage λ. Does not carry the λ prophage.

This genotype tells us that the strain carries no plasmids or phage λ and so can be used for plasmid transformation or assay of phage. It can be transformed efficiently by plasmid DNA and then used for the production of large-scale preparations of plasmid DNA. It can grow on glucose but is unable to grow on a range of sugars as sole sources of carbon and energy. A chemically defined medium would need to have proline added, but this would normally be supplied by peptone in complex media. Although this strain is Lac⁻ it does carry a wild-type gene for β-galactosidase (*lacZ*⁺) so this strain is unsuitable for use as a host for plasmids which depend on lactose fermentation to give a blue coloration with X-gal (Section 3.5). The *supE* locus means that it will suppress amber mutations, for example in phage λ, thus allowing it to produce plaques.

Space does not permit a complete glossary of a wide range of symbols and the reader is referred to *Appendix 1* and to the *Molecular Biology Labfax* (11). The latter contains genotypes of all relevant strains as well as a list of gene symbols.

3.2 Characterization of nutritional mutants

It is normally only essential to check for genetic characters which you intend to use in a particular experiment. Thus if absence of β-galactosidase activity is essential then you should check that the strain is Lac⁻. Similarly if you needed to select for the Pro⁺ phenotype then you would need to check that the strain is proline requiring. A method illustrating the general points is given in *Protocol 5*. However, if you suspect that your strain has mutated, reverted or is a contaminant, it may by necessary to do a more thorough check of other markers using the same general approach.

Protocol 5

Analysis of nutritional mutants

Reagents

• PBS or Ringer's solution (see *Protocol 2*)

Method

1 Prepare a stock solution of a chemically defined minimal agar medium lacking a carbon and energy supply or growth factors.

2 Examine the genotype of the strain(s) for analysis and identify the growth factors to be analysed. The system will be illustrated for HB101 and the *pro*, *lac*, and *mtl* markers only.

3 Label four tubes (20 ml capacity) with the letters A, B, C, and D and make additions as indicated:

	A	B	C	D
proline	+	+	+	−
glucose	+	−	−	+
lactose	−	+	−	−
mannitol	−	−	+	−

Use a 1% (w/v) proline stock solution and dilute 1 in 100; use 20% (w/v) sugar stocks diluted to 0.1% final volume.

4 Add the contents of each tube to 20 ml agar medium and pour four plates (see *Protocol 1*).

Protocol 5 continued

5   Suspend single colonies of HB101 and a wild-type control in 0.5 ml PBS or dilute Ringer's solution and vortex.

6   Streak a loopful of HB101 and the wild-type control on to all four media at previously marked positions.

7   Incubate for 16–24 h and record growth. Always compare the relative growth of the mutant strain on the media with the growth of the wild-type strain.

8   Analysis of mutants defective in the utilization of different nitrogen, sulphur or phosphorus sources requires a minimal medium lacking these supplements.

3.3 Characterization of antibiotic resistance

The resistance of bacterial strains to antibiotics is a very useful selective technique. Genes controlling this resistance can be carried on the bacterial chromosome or on a plasmid and the latter are very useful in ensuring that a particular plasmid is present (see Chapter 8, Section 3.1). However, it should be remembered that plating large numbers of sensitive cells on media containing an antibiotic can select rare spontaneously-resistant mutants due to chromosomal mutation rather than to the receipt of a resistant plasmid. Similarly, contaminants may also be naturally resistant to the same antibiotic and even a low level of contamination will be readily revealed on the selective plates. Because antibiotics are often thermolabile they cannot be sterilized by autoclaving, therefore solutions should be membrane filtered. Most antibiotics are supplied sterile and it is possible to make up solutions using aseptic technique in sterile distilled water. The actual concentration of antibiotic required depends on the conditions used. In general, higher concentrations are required for high density of cells on agar plates while low concentrations are needed for low densities of cells growing in liquid medium. Detailed techniques for checking antibiotic resistance are shown in *Protocol 6* and *Appendix 2* lists some of the points to note about particular antibiotics.

Protocol 6

Analysis of antibiotic resistance

Reagents

- Stock solutions of antibiotics (see *Appendix 2*)
- LB agar plates (see *Protocol 3*)
- PBS or Ringer's solution (see *Protocol 2*)

Method

1   Make up stock solutions of antibiotics as described in *Appendix 2*, using aseptic technique and sterile distilled water. Filter through a sterile membrane, pore size 0.45 μm, if desired.

Protocol 6 continued

2 Distribute in aliquots of 200 μl into labelled 1.5-ml microfuge tubes and store at −20°C until required.

3 Make up a suitable nutrient medium such as LB agar and cool to 50°C. Pour and dry a few plates (see *Protocol 1*) without the antibiotic, to act as controls.

4 Add the antibiotic to the remainder of the nutrient medium to give the desired concentration and mix well.

5 Pour and dry the antibiotic plates. Most antibiotic plates will keep for at least 2–3 weeks.

6 Mark each set of agar plates with a suitable grid and label with the strains to be tested. It is best to include control strains which are sensitive to the antibiotic to test the efficiency of the antibiotic activity.

7 Suspend a single colony of a strain to be tested in 200 μl of PBS or dilute Ringer's. Vortex to suspend the cells.

8 Flame and cool a Nichrome loop about 3–4 mm diameter and remove a loopful of cell suspension.

9 Streak this suspension over a distance of 1–2 cm on each antibiotic plate. Repeat on the medium without antibiotic.

10 Repeat steps 7–9 with each strain to be tested, including the control strain.

11 Incubate at 37°C for 16–36 h depending on the strain and the antibiotic. Compare the growth of the strains on the antibiotic and control media. Note that if antibiotic activity is low or absent (due possibly to inactivation because the agar was not cooled sufficiently in step 3) then the control strain will grow on both media.

12 For large-scale checking of transformants or other recombinants, it is more convenient to pick off individual colonies with sterile toothpicks and to streak plates with these directly.

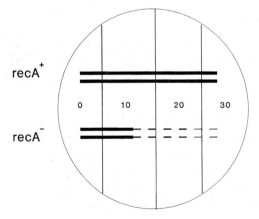

Figure 4 Procedure for characterizing *recA* strains. The marked areas show the expected growth after irradiation and incubation.

3.4 Characterization of *recA* and UV-sensitive mutants

Recombination-deficient mutants, such as *recA*, are used to minimize recombination between cloned genes and homologous regions on the chromosome. In addition, the use of *recBC* mutants can minimize the loss of repetitive DNA in gene libraries. It is difficult to check for recombination deficiency, but luckily there is an associated phenotype of UV-sensitivity which is easy to analyse. As with any genetic character, it is important to have both *recA*⁻ and *recA*⁺ strains so that a comparison can be made. The technique is as given in *Protocol 7*.

Protocol 7

Checking strains for the Rec⁻ phenotype (see *Figure 4*)

Equipment and reagents

- LB agar plates (see *Protocol 3*)
- Bactericidal UV lamp (254 nm)

Method

1 Prepare overnight broth cultures from single colonies of Rec⁺ and Rec⁻ cultures.

2 Using a flamed Nichrome loop, streak each strain across a well-dried LB plate such that the streak is about 6 cm. Allow the liquid to dry.

3 Mark the base of the agar plate so that each streak is divided into four segments labelled, left to right, 0, 10, 20 and 30 s.

4 Cover the plate with a piece of card such that only the 30 s segment is uncovered. Remove the lid of the agar plate and expose to a bactericidal UV lamp (254 nm) at a height of about 30 cm for 10 s. **Caution: wear UV safety goggles and gloves during these operations.**

5 Move the card across so that the 20 s segment is exposed for 10 s.

6 Move the card across so that the 10 s segment is exposed for 10 s.

7 Switch off the lamp, wrap the plate rapidly in foil to avoid light-induced repair processes and incubate at 37 °C overnight.

8 The segments will have received 0, 10, 20 or 30 s exposure to UV and there should be a clear distinction between the two phenotypes.

3.5 Characterization of the utilization of lactose: X-gal

There are several ways of detecting the utilization of sugars such as lactose. One method makes use of a chemically defined medium in which lactose is the sole source of carbon and energy. It is used to replace glucose and, in this medium, only Lac⁺ strains will be able to grow. Alternatively, a nutrient medium with lactose can be used which contains the indicator dyes eosin and methylene blue. When lactose is fermented, acid is produced and the indicators change colour: Lac⁺ colonies with β-galactosidase activity give dark purple colonies with a green fluorescent sheen whereas Lac⁻ colonies are pink. A more expensive way of analysing for this characteristic is to use the substituted β-galactoside sugar X-gal. Media are normally prepared containing this compound, together with IPTG which fully induces the *lacZ* gene but which is not a substrate for the enzyme. Details are given in Chapter 8, Section 3.2.

3.6 Detection of lysogeny

Bacteria carrying the prophage of temperate phages such as λ and P2 have been exploited in a variety of ways. The λ lysogens have been used to produce packaging extracts for isolation of genomic DNA libraries and P2 lysogens are important in the Spi⁺ selective system for some types of recombinant λ phages (Chapter 8, Section 3.4). When it is necessary to check for lysogeny it is essential to have a strain which is sensitive to the phage. Thus, for a (P2)⁺ lysogen it is necessary to have a second strain which is sensitive to phage P2 as well as, ideally, a sample of phage P2 itself. The simplest check is to make a single streak of each strain, such as *E. coli* Q359 (P2)⁺ and Q358 (P2)⁻, on an LB agar plate. The liquid is allowed to dry and a small volume of a P2 stock is spotted on the streak. Q359 is immune to the phage due the presence of a repressor, while Q358 is lysed. If you do not have any P2 phage then prepare pour-plates of each strain and spot overnight broth cultures of Q358 and Q359 on to each lawn. Q359 produces low numbers of free phages, owing to spontaneous lysis, and these appear as a thin halo of lysis round the spot of Q359 on the lawn of Q358 (*Figure 5*). Similar techniques will also work for λ phage and lysogens.

3.7 Screening for plasmids

It is essential to have a good accurate map of the plasmid with details of its genotype. The most popular plasmid vectors are described in Chapter 9, and detailed compilations have been published (11). The easiest characters to detect are antibiotic resistance. For example, the pUC and pGEM series of vectors are resistant to ampicillin, so a simple check can therefore be made by streaking on

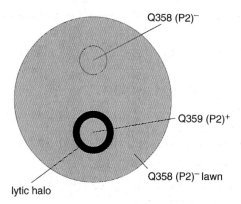

Q358 (P2)⁻

Q359 (P2)⁺

Q358 (P2)⁻ lawn

lytic halo

Figure 5 Detection of lysogeny by spot testing of strains on a bacterial lawn.

media containing ampicillin. It is advisable to include the selective antibiotic, whether this is ampicillin or some other compound, in all media used for the growth and maintenance of plasmid-containing bacteria because the *par* locus, which controls regular partition, has been deleted in most plasmids. Cells lacking plasmids can therefore be generated when cells divide but these will not survive if the selective agent is present in the medium. Further preliminary screening depends on other markers carried by the plasmid, and final checking will depend on extraction, restriction and electrophoresis of the plasmid DNA.

3.8 Microscopic examination of cultures

A visual check of a culture may suggest contamination, or it may be that a rare transformant simply does not look right. Rather than extracting and analysing plasmid DNA it may be worth having a look at the organism to check that it is actually Gram-negative and rod-shaped. It is even quicker to do a simple smear followed by a methylene blue stain as this will quickly eliminate budding yeasts and cocci. *Protocol 8* describes the basic methods.

Protocol 8

Techniques for the microscopic examination of strains

Equipment and reagents

- Loeffler's methylene blue, crystal violet, Lugol's iodine, ethanol and dilute fuchsin
- Microscope with oil immersion lens, plus microscope slides

Method

1 Clean a microscope slide by immersing it in alcohol and flaming off in a Bunsen burner.

2 Either place a loopful of a cell suspension onto the slide or suspend cells from an agar plate in a small drop of water on the slide.

3 Spread the suspension thinly and evenly over the slide. This is best done with the edge of a second microscope slide. The smear should be only just visible and the aim is to produce a monolayer of cells.

4 Leave the slide to air-dry or hold high over a Bunsen burner to dry the smear slowly.

5 Fix the cells by passing the dried slide slowly three times through the Bunsen flame. The bacteria will adhere to the slide.

6 Place the smear on a staining rack (two glass tubes held together by short pieces of rubber tubing). Flood the slide with methylene blue and leave for 5 min.

7 Wash the slide under the tap and blot dry with blotting or filter paper.

8 Add a small drop of immersion oil and view down the microscope using the oil immersion lens. Look for areas where the cells are well spread out.

Protocol 8 continued

9 Alternatively, use the Gram stain:

(a) Stain in crystal violet for 1 min, then wash well under running water and blot dry.

(b) Wash with Lugol's iodine and then stain for 1 min with fresh Lugol's iodine.

(c) Wash with tap water and blot dry. Decolourize with ethanol until no more dye is washed out.

(d) Wash with tap water and counterstain with dilute fuschin for 30 s.

(e) Wash and blot dry. Gram-negative cells will appear red whereas Gram-positive cells will be dark purple. It is best to have examples of both types when you are practising the technique.

4 Preservation of stock cultures

A culture which has been checked and used successfully is a valuable laboratory asset and should be preserved as efficiently as possible. It is sensible to preserve the culture by a number of methods as you will require the strain for short-term and long-term use. The broth and plate cultures already described can be used for several weeks or even months, but there are problems inherent in using cultures stored at 4°C. Cultures may become contaminated or lose viability; they can also mutate or be infected by phages or other plasmids. Some micro-biologists believe that the only stable strain is a dead strain. This is true, but we can get close to complete lack of growth and slow chemical change in frozen cultures. Alternatively, freeze-dried cultures are convenient to keep and are not susceptible to damage following electrical breakdown. There are several good reviews of preservation methods (12, 13), therefore detailed techniques on freeze-drying will not be given here.

4.1 Preservation of short-term cultures

Broth cultures of *E. coli* will retain viability for several weeks if stored in sealed containers at 4°C. This can be recommended for cultures which lack plasmids and show little variation, such as *E. coli* HB101. The use of broth cultures older than a few days for strains carrying unstable plasmids is not recommended as the antibiotics used to select the plasmid may have been inactivated and growth of cells without the plasmid may have occurred. It is much safer to use an agar plate with single colonies which have previously been checked. The major problems with agar plates are desiccation and contamination. These problems can be reduced by sealing the edge of the plates with parafilm or adhesive tape. Plates like this can be used successfully for 2–3 months, but for particularly critical experiments you should check your strains again by re-streaking on a selective or diagnostic medium.

step 9. This gives an approximate titre which is used to make a full assay by suspending a known phage concentration with the bacterial suspension in the overlay. When the titre is known in advance a limited range of dilutions can be prepared. The details of these techniques are given in *Protocol 10*.

Protocol 10

Assay of phages by the plaque method

Reagents

- SM (λ storage and dilution medium: 5.8 g NaCl, 2 g MgSO$_4$.7H$_2$O, 50 ml 1 M Tris-HCl, pH 7.5, 5 ml 2% [w/v] gelatin solution, water to make 1 l). Sterilize by autoclaving
- DYT agar plates (16 g bacto-tryptone, 10 g bacto-yeast extract, 5 g NaCl, 15 g bacto-agar per 1 l). Check the pH and adjust to 7.0–7.2 with NaOH; sterilize by autoclaving

- LTS top agar (same recipe as DYT except with 6 g agar/l)

Method

1 Add 1 ml of SM to a series of sterile plastic vials. Label the tubes –2, –4, –6, –8 and –10.

2 Add 10 μl of phage stock to tube –2 and mix well. This dilution is 10^{-2}. Repeat this operation to produce the dilution series.

3 Prepare a thick agar plate of a rich medium such as DYT on a level area of bench and dry the plate well. Warm the plate by incubating at 37 °C.

4 Prepare 3 ml of LTS top agar, melt, and maintain at 50 °C.

5 Add 200 μl of an overnight broth culture[a] of a suitable host strain to the molten agar at 50 °C.

6 Mix the bacteria and the agar. Wipe the outside of the tube and pour rapidly over the DYT plate (again on a even surface). Rock and swirl the plate to spread the agar evenly over the surface (this must be done immediately).

7 Allow the agar to set, and dry the plate for about 15 min at 45 °C.

8 At marked places, spot 20 μl of the six dilutions and the neat suspension on the overlay plate. Allow the liquid to dry before inverting the plate. Incubate at 37 °C overnight.

9 Count the plaques where there are 5–20 per drop. The titre of the phage/ml is 50 times the count multiplied by the dilution factor.

10 For a more accurate assay, or when a complete plate at the same dilution is required:

   (a) Prepare as many nutrient plates as necessary. If large plaques are required for assay or production of stocks use thick plates (40 ml) while for screening use thin plates (20 ml).

   (b) Calculate the required volume of diluted phage for between 30 and 300 phage particles (or 1000+ for screening) and add this to 3 ml of molten agar medium at 50 °C as described in step 4.

Protocol 10 continued

(c) Add 200 μl of indicator bacteria and mix bacteria, phage and agar.

(d) Pour quickly over the plate as before, allow to set and incubate at 37°C overnight. Count the plaques as before.

[a] Bacterial cultures for λ should be grown on a medium containing 0.2% (w/v) maltose (see *Appendix 2*) to induce receptors. The LTS agar should also contain 0.2% maltose. The broth culture can be stored at 4°C for up to 1 month. For M13, male bacterial strains (Hfr or F^+) must be used; better results will be obtained if the lawns are prepared from actively-growing cultures.

5.4 Purification of phage by single-plaque isolation

All phage stocks should be purified periodically by single-plaque isolation and checked for appropriate genotypes or DNA restriction patterns. This is equivalent to single-colony isolation for bacteria (see Section 2.1). The same technique is used to purify a phage containing cloned DNA. Where the purity of a stock is critical it is best to choose well-separated plaques on fresh plates, as phages may diffuse sideways in the agar. A second purification step will avoid these problems. Stab an appropriate plaque, prepared by the technique in *Protocol 10*, with the tip of a sterile Pasteur pipette and rock from side to side so that a plug of agar containing the plaque is removed in the pipette tip. With a rubber teat or pipette dispenser, blow the plug into 1 ml of SM (*Protocol 10*) in a sterile plastic tube. Leave for 1 h so that the phage diffuses into the SM. An average plaque will contain about 10^6 phages. Plate a suitable dilution of the phage suspension to obtain single plaques and then prepare a fresh phage stock by one of the methods detailed below.

5.5 Preparation of small-scale phage stocks

The plate method depends on producing almost confluent lysis on a plaque assay plate. It has the advantage that it is reliable and predictable but the disadvantage that agar has to be removed from the preparation. Agar contains inhibitors which can interfere with restriction, ligation, and other enzymes involved in DNA manipulations. If this is found to be a major problem it is better to replace the agar with agarose.

Protocol 11

The plate method for preparing a λ phage stock

Reagents

- SM (see *Protocol 10*)
- Chloroform

Method

1 Prepare a series of up to five agar plates displaying almost confluent lysis by the method described in *Protocol 10*, step 10. The number of phage particles required will vary depending on the size of the plaques but normally will be between 5000

Protocol 11 continued

and 100 000. Plaques should just be visible in isolated areas. Plates with complete lysis are less reliable as lysis without phage replication may have occurred.

2   If mutant plaques are likely to be a problem (e.g. clear plaque mutants in a wild-type stock) then use the minimum number of phages required for almost confluent lysis and process the plates at either 6 h or about 15 h.

3   Add 5 ml of SM to each plate and scrape the 0.6% agar off the plates into a sterile 250-ml conical flask with a sterile glass spreader.

4   Add 0.5 ml of chloroform, vortex the flask for 1 min and leave to stand for 1 h so that the phage can diffuse from the agar.

5   Decant the mixture into a centrifuge tube (resistant to chloroform) and centrifuge at 5000 g for 10 min to bring down bacterial debris and agar fragments.

6   Remove the supernatant carefully and filter to sterilize or add to a sterile plastic tube over a few drops of chloroform.

7   Assay the stock by one of the methods described in *Protocol 10*.

Notes

(a) The method can be modified by using agarose instead of agar or by allowing the phage to diffuse into the SM by gentle shaking of the intact plate for 3 h. Both of these methods will reduce the contamination of the stock by inhibitors.

(b) The method is useful when rapid lysis mutants or clear plaque mutants are a potential problem as diffusion is restricted by the agar and mutants will not overgrow the required phage, as can happen in broth culture methods.

The broth method is cleaner because it is only necessary to remove bacterial debris, but it is less predictable because it is essential to have the correct value for m to give maximum lysis (see Section 5.2). If a stock culture is required quickly then you will need to set up a series of cultures with different ratios of phage and bacteria.

Protocol 12

The broth method for preparing λ phage stocks

Reagents

- LB broth (see *Protocol 9*). For this procedure add $CaCl_2$ to a final concentration of 5 mM from a filter-sterilized stock solution immediately before use

- Chloroform

Method

1   This method can be used on any scale from 5 ml to 500 ml cultures. For λ use LB broth containing 5 mM $CaCl_2$ but no $MgCl_2/MgSO_4$ or maltose. The method can also work with more complex media.

47

5.8 Induction of λ lysogens

Traditionally, lysogenic cultures were induced by UV irradiation. Damage to the DNA occurred followed by derepression of the *recA* gene product which then destroyed the λ *cI* repressor. The use of UV is not recommended as it can induce mutations in phage particles, and most lysogens are now induced by a temperature-sensitive system. With many λ strains the repressor protein is thermolabile at 45 °C and thus lysogenic cultures can be induced to undergo the lytic cycle by a brief treatment at this temperature. The S^+ gene product is required for natural lysis and some phage strains have amber mutations or chain-terminating triplets in this gene. If the host strain has the equivalent suppressor then natural lysis will occur, otherwise it has to be induced with chloroform.

Protocol 14

Induction of λ lysogens

Equipment and reagents

- Shaking water baths at 37°C and 45°C
- LB agar plates (see *Protocol 3*)
- LB broth (see *Protocol 9*)
- SM (see *Protocol 10*)
- Chloroform

Method

1 Prepare duplicate streak plates of the lysogenic culture on LB plates and incubate at 30 °C and 42 °C. No growth should occur at 42 °C owing to induction of the lytic cycle.

2 Use two shaking water baths at 37 °C and 45 °C, respectively. Grow an overnight 20 ml LB broth culture at 37 °C and inoculate into 400 ml of LB in a 2-l flask. Shake at 250 r.p.m. at 37 °C until the OD at 550 nm reaches 0.6 (about 3×10^8 cells per ml).

3 Rapidly increase the temperature in the broth culture to 45 °C. This is best done in a large volume of water at 70 °C. Monitor the temperature, and when this reaches 45 °C, transfer the flask to the water bath at 45 °C. Incubate with vigorous shaking for 15 min.

4 Transfer the flask to the 37 °C water bath and continue shaking.

5 Monitor for lysis (if suppressor mutations are present), as in *Protocol 12*, step 5. Samples can be withdrawn and tested for lysis with a small volume (0.05 volume) of chloroform. Vigorous shaking should be followed by lysis seen as translucence or rope-like debris.

6 Strains which do not suppress amber mutations will not lyse spontaneously. After about 2 h at 37 °C, collect the cells by centrifugation at 3000 **g** and suspend in 10 ml of SM. Add 500 μl of chloroform and vortex vigorously.

7 Stand the suspension for up to 1 h at room temperature.

8 Centrifuge to remove cell debris and treat as for standard λ phage stock, as described in *Protocols 12* and *13*.

Protocol 11 continued

and 100 000. Plaques should just be visible in isolated areas. Plates with complete lysis are less reliable as lysis without phage replication may have occurred.

2   If mutant plaques are likely to be a problem (e.g. clear plaque mutants in a wild-type stock) then use the minimum number of phages required for almost confluent lysis and process the plates at either 6 h or about 15 h.

3   Add 5 ml of SM to each plate and scrape the 0.6% agar off the plates into a sterile 250-ml conical flask with a sterile glass spreader.

4   Add 0.5 ml of chloroform, vortex the flask for 1 min and leave to stand for 1 h so that the phage can diffuse from the agar.

5   Decant the mixture into a centrifuge tube (resistant to chloroform) and centrifuge at 5000 g for 10 min to bring down bacterial debris and agar fragments.

6   Remove the supernatant carefully and filter to sterilize or add to a sterile plastic tube over a few drops of chloroform.

7   Assay the stock by one of the methods described in *Protocol 10*.

Notes

(a) The method can be modified by using agarose instead of agar or by allowing the phage to diffuse into the SM by gentle shaking of the intact plate for 3 h. Both of these methods will reduce the contamination of the stock by inhibitors.

(b) The method is useful when rapid lysis mutants or clear plaque mutants are a potential problem as diffusion is restricted by the agar and mutants will not overgrow the required phage, as can happen in broth culture methods.

The broth method is cleaner because it is only necessary to remove bacterial debris, but it is less predictable because it is essential to have the correct value for m to give maximum lysis (see Section 5.2). If a stock culture is required quickly then you will need to set up a series of cultures with different ratios of phage and bacteria.

Protocol 12

The broth method for preparing λ phage stocks

Reagents

• LB broth (see *Protocol 9*). For this procedure add $CaCl_2$ to a final concentration of 5 mM from a filter-sterilized stock solution immediately before use

• Chloroform

Method

1   This method can be used on any scale from 5 ml to 500 ml cultures. For λ use LB broth containing 5 mM $CaCl_2$ but no $MgCl_2$/$MgSO_4$ or maltose. The method can also work with more complex media.

Protocol 12 continued

2 Grow an overnight culture of a suitable host and add 100 μl to 5 ml in a 20 ml tube, or 10 ml to 500 ml in a 2-l conical flask.

3 Incubate at 37 °C at 250 r.p.m. until the OD at 550 nm is about 0.2. This should be approximately 10^8 exponentially growing cells/ml. Incubation should take about 2–3 h.

4 Add phage to give an m of 0.01. For the 5 ml culture this means adding the phage from two or three plaques (removed as agar plugs; see Section 5.4), or for the 500 ml culture the phage from one or two plates showing almost confluent lysis (prepared by the method in *Protocol 11*). The latter can be done simply by scraping the agar overlay into the flask. Alternatively, prepare a phage stock from the plates and add the agar-free stock directly. It is important to remove all traces of chloroform by aeration before adding.

5 Continue shaking until the culture lyses, which will normally take between 2 and 4 h. This can be detected by a rapid drop in OD accompanied by the production of 'ropes' of lysed bacteria. If this does not occur after 5 h, either continue incubation overnight or add chloroform (200 μl per 5 ml culture or 20 ml per 500 ml culture). Continue shaking until lysis occurs.

6 The production of efficient lysis is essential for a high phage yield. The presence of clearing and 'rope' formation is a reliable indication of a high titre stock. If this does not occur it is better to repeat the procedure until good lysis is detected.

7 If not already added, add chloroform to lysed cultures as in step 5 above. Shake for 10 min and then harvest the cultures into centrifuge tubes which will resist chloroform. Centrifuge for 15 min at 11 000 g to remove cell debris and agar.

8 Remove the supernatant and store with a few drops of chloroform at 4 °C.

5.6 Preparation of large-scale phage stocks

As described in *Protocol 12*, the broth method for preparation of small-scale phage stocks can be scaled up to 500 ml cultures in 2-l flasks.

5.7 Purification of λ phage particles

For many purposes, phage preparations need to be free from bacterial nucleic acids and other contaminants such as bacterial carbohydrates or agar. Bacterial DNA can be incorporated into gene libraries and other contaminants can interfere with ligase or restriction enzyme activities. Intact phage particles are resistant to nucleases, so a treatment with RNase and DNase can remove these contaminants. Phage particles can be purified by differential centrifugation but damage to the particles occurs. PEG has been used effectively to precipitate the phage particles making centrifugation easier. The most efficient way of purifying the phage particles is on a CsCl gradient, as this separates the particles from carbohydrates. A method for purifying λ phage is given in *Protocol 13*. Extraction of DNA from these particles is straightforward (see Chapter 3, Section 2.2).

Protocol 13

Purification of phage λ

Reagents

- DNase I (stock solution of 1 mg/ml prepared in 20 mM Tris-HCl, pH 7.5, 1 mM MgCl$_2$, stored at $-80\,°C$)
- RNase I (stock solution of 10 mg/ml prepared in 10 mM Tris-HCl, pH 7.5, 15 mM NaCl)

- PEG 6000
- SM (see Protocol 10)
- Chloroform

Method

1 Prepare a high-titre stock (at least 10^{10}/ml) by the method described in *Protocol 12*. Measure the volume and add 1 μg/ml of both DNase I and RNase I to remove bacterial nucleic acids. Incubate at room temperature for 30 min.

2 Add 40 g solid NaCl per litre and dissolve by gentle agitation.

3 Add 140 g PEG 6000 per litre. Add this slowly with constant gentle mixing on a magnetic stirrer at room temperature.

4 Leave at 4°C overnight to precipitate the phage.

5 Swirl the flask gently to resuspend any sediment.

6 Centrifuge at 11 000 g for 10 min to collect the precipitated phage.

7 Decant the supernatant and discard. Invert the centrifuge tube and drain well. Wipe the inside of the tube with a paper tissue.

8 Resuspend the pellet in 16 ml of SM (per litre of original stock). Add an equal volume of chloroform and stir on a magnetic shaker at 37°C for 1–2 h until the pellet is fully resuspended. Care should be taken at this stage as losses can occur.

9 Centrifuge at 2000 g to separate the phases and carefully remove the upper aqueous layer containing the phage, which can be stored as a stock.

10 Losses can occur at a number of stages and it is useful to do a Miles–Misra assay (see *Protocol 3*, step 9) for phage at various steps such as original stock, the supernatant after PEG precipitation and the final stock. Supernatants should be retained until the results are known so that further recovery can be made and phage stocks pooled.

11 Add 0.75 g CsCl per ml of stock and dissolve by inversion. This should give a density of 1.5 g/ml. Centrifuge at 100 000 g for 24 h and phage should produce a bluish band half-way down the tube. Remove the band with a 21-gauge syringe needle.

12 Dialyse against 2 l of SM overnight and store at 4°C with a few drops of chloroform.

5.8 Induction of λ lysogens

Traditionally, lysogenic cultures were induced by UV irradiation. Damage to the DNA occurred followed by derepression of the *recA* gene product which then destroyed the λ *cI* repressor. The use of UV is not recommended as it can induce mutations in phage particles, and most lysogens are now induced by a temperature-sensitive system. With many λ strains the repressor protein is thermolabile at 45 °C and thus lysogenic cultures can be induced to undergo the lytic cycle by a brief treatment at this temperature. The S^+ gene product is required for natural lysis and some phage strains have amber mutations or chain-terminating triplets in this gene. If the host strain has the equivalent suppressor then natural lysis will occur, otherwise it has to be induced with chloroform.

Protocol 14

Induction of λ lysogens

Equipment and reagents

- Shaking water baths at 37°C and 45°C
- LB agar plates (see *Protocol 3*)
- LB broth (see *Protocol 9*)
- SM (see *Protocol 10*)
- Chloroform

Method

1. Prepare duplicate streak plates of the lysogenic culture on LB plates and incubate at 30 °C and 42 °C. No growth should occur at 42 °C owing to induction of the lytic cycle.

2. Use two shaking water baths at 37 °C and 45 °C, respectively. Grow an overnight 20 ml LB broth culture at 37 °C and inoculate into 400 ml of LB in a 2-l flask. Shake at 250 r.p.m. at 37 °C until the OD at 550 nm reaches 0.6 (about 3×10^8 cells per ml).

3. Rapidly increase the temperature in the broth culture to 45 °C. This is best done in a large volume of water at 70 °C. Monitor the temperature, and when this reaches 45 °C, transfer the flask to the water bath at 45 °C. Incubate with vigorous shaking for 15 min.

4. Transfer the flask to the 37 °C water bath and continue shaking.

5. Monitor for lysis (if suppressor mutations are present), as in *Protocol 12*, step 5. Samples can be withdrawn and tested for lysis with a small volume (0.05 volume) of chloroform. Vigorous shaking should be followed by lysis seen as translucence or rope-like debris.

6. Strains which do not suppress amber mutations will not lyse spontaneously. After about 2 h at 37 °C, collect the cells by centrifugation at 3000 **g** and suspend in 10 ml of SM. Add 500 μl of chloroform and vortex vigorously.

7. Stand the suspension for up to 1 h at room temperature.

8. Centrifuge to remove cell debris and treat as for standard λ phage stock, as described in *Protocols 12* and *13*.

5.9 Techniques involving phage M13

Many of the techniques which apply to phage λ also apply to M13. However, a few points should be noted. The bacterial host must be a male strain containing the F factor as the phage infects via the sex pilus (an appendage coded by the sex factor). Plaques can be produced either by transforming competent *E. coli* cells with the double-stranded replicative form of the DNA (Chapter 8, Sections 2.1 and 2.2) or by infecting cells with intact phage particles containing a single-stranded circular molecule. Since phage particles do not lyse the host, it is possible to prepare stocks of M13 simply by inoculating a broth culture of an F⁺ strain with phage particles and growing to stationary phase. The bacteria are removed by centrifugation at $10\,000\,g$ for 15 min. The supernatant contains free phage particles released through the wall of the host. The phage can be purified by a PEG precipitation method similar to *Protocol 13*, as described in Volume II, Chapter 6.

5.10 Methods for preserving phage stocks

The simplest method for keeping a phage stock is to seal a Petri dish containing separate plaques of a purified phage type. Parafilm or tape is used to seal the edges of the plate, which is kept at 4 °C. Viable phage can be isolated from single plaques for several months, but this method is not recommended if there is a variety of different phage types in the same area, since cross-diffusion may occur. Once a high-titre stock has been prepared in SM this can be kept over a small volume of chloroform at 4 °C for many months. The chloroform may evaporate with time and should be replenished or the stock may become contaminated with fungi or bacteria. Stocks can also be preserved by the addition of 7% (v/v) DMSO. Add DMSO and mix gently, then freeze rapidly in liquid nitrogen or a dry-ice–alcohol mix and store at −70 °C. Scrape the surface with a sterile Nichrome loop and spot on to a overlay lawn of a sensitive host to test for viability.

6 Troubleshooting

6.1 General principles

Techniques are always easy when you know how or when you can watch an expert using the method. However, it is not uncommon to find that the technique is simply not working. The speed of solving these technical problems makes a big impact on the time available for research and rate of progress. One basic principle is to include controls whenever possible, since these can show clearly what has gone wrong. Solving more difficult problems is very much like detective work—following up clues, eliminating possible suspects and testing theories. It makes sense to ask others who have managed to get the technique to work and to show them your plates or photographs of puzzling results. In microbiology, you should be fully aware of where microorganisms can grow: they can grow wherever conditions are not inhibitory. Algae and bacteria have been found growing in distilled water, and phosphate buffer with a low glucose concentration is an ideal growth medium. Solutions which were sterile last week may have been opened and left on the bench, and be heavily contaminated. If they are, you can be sure that the contaminant produces a nuclease which will destroy your precious DNA.

6.2 Contamination

6.2.1 Contaminated cultures

The source of contaminated cultures may be the air, the apparatus or the growth medium. The culture may have been contaminated when it was received. A visual inspection of a streak plate should show the most obvious contaminants and these can also be checked by their phenotypes and microscopic properties. More subtle changes are difficult to check and the failure of a technique to work may be the only clue.

If you are not sure which colony on a plate is the correct strain you may have to prepare a range of cultures and carry out some simple tests to see which are, for example, sensitive to λ phage or which can conjugate with a known Hfr strain. Genetic variation is the most difficult to detect and may not become apparent for some time. Chromosomal rearrangements such as transposition may only become apparent on genetic mapping or on analysis of antibiotic resistance. Gene mutation or reversion will only be detected if you check for these particular genetic markers. If all else fails and the technique is still not working, ask someone else for an alternative culture or order a substitute from a commercial supplier.

6.2.2 Contaminated media, solutions and equipment

Your strains may be correct but you may be contaminating them every time you attempt to perform an experiment. The inclusion of uninoculated controls should indicate the source of the problem. Media with disc-shaped colonies probably means that one of the media components is contaminated. If you suspect that solutions and buffers are contaminated then spread them on nutrient media and look for colonies after incubation. Wherever possible, solutions should be sterilized, stored at 4 °C and opened using sterile technique. If a solution is used frequently it is much better to sterilize it in suitable aliquots and discard material once it has been opened a few times. Contaminated equipment implies that your quality control on sterilization or storage is inadequate. Autoclave tape may indicate that your use of the autoclave is at fault or it could be that the environment is contaminating the equipment after sterilization. Glass pipettes and other glassware generally can be flamed before use, but this is not possible for plasticware previously sterilized by gamma radiation. If a particular piece of apparatus, such as a centrifuge tube, is suspected then add some sterile nutrient medium, incubate and look for turbidity.

6.3 Poor growth of bacteria or phage

Techniques often quote a time-course for a particular experiment; if your cultures are growing much slower than this then you may suspect that there are factors limiting growth. The first component to suspect is that the growth medium is inadequate. A check of the genotype of the bacterial strain should reveal if there are growth factor requirements which may be absent from the medium; for example, peptone may be deficient in arginine and/or lyrine, and vitamins may be absent from yeast extract. Alternatively, the concentration of an antibiotic may be too high and inhibition may have occurred. Components of the medium can be altered and the growth rate of the strain checked again. A low level of contamination can also affect growth, particularly if the contaminant is a streptomycete that produces an antibiotic which inhibits the strain under study. Another factor which affects growth is the aeration of the culture as large volumes of media with inadequate shaking can rapidly become oxygen-limited. Ideally, a flask of 250 ml capacity should have no more than 25 ml of medium to avoid oxygen-limitation. In practice, the earlier phases of growth are less likely to be oxygen-limited and larger volumes can be used without affecting doubling times. A common source of problems is inhibition by chemicals contaminating the glassware. Glassware is often used many times and may have been washed in a dish-washer that has left residues of salts and detergents. These can rapidly destroy phage λ and, consequently, plastic tubes are best for the handling of phages. You are strongly advised to wash your own glassware and to make sure that rinse cycles are sufficient to remove these chemical contaminants. Poor growth of phages can be due to a variety of factors such as resistance of the host to phage infection and absence of cofactors essential for adsorption and infection. Sudden lysis of cultures may be due to the presence of air-stable phages in the environment. A rich medium with a good carbon source will obviously give better bacterial growth than a chemically defined medium where the organism must synthesize almost everything for itself. Good bacterial growth will normally give good phage growth but note should be taken of the optimum conditions for handling λ, as discussed in Section 5.

References

1. Prescott, L. M., Harley, J. P., and Klein, D. A. (1999). *Microbiology*, 4th edn. WCB/McGraw-Hill, Boston.
2. Bainbridge, B. W. (1987). *Genetics of microbes*. Blackie, Glasgow.
3. Health and Safety Executive (1996). *A guide to the genetically modified organisms (contained use) regulations 1992, as amended in 1996*. HMSO, London.
4. Anonymous (1984). *Federal Register*, 49, Part VI, Number 227, 46266.
5. Collins, C. H. (ed.) (1988). *Safety in clinical and biomedical laboratories*. Chapman and Hall, London.
6. Collins, C. H. (ed.) (1999). *Laboratory-acquired infections: history, incidence, causes and prevention*, 4th edn. Butterworth-Heinemann, Oxford.
7. Health and Safety Commission (1993). *Control of substances hazardous to health; and control of carcinogenic substances*, 4th edn. HMSO, London.
8. Clowes, R. C. and Hayes, W. (1968). *Experiments in microbial genetics*. Blackwell, Oxford.
9. Miller, J. H. (1972). *Experiments in molecular genetics*. Cold Spring Harbor Laboratory Press, Cold Spring Harbor.
10. Demerec, M., Adelberg, E. A., Clark, A. J., and Hartman, P. E. (1966). *Genetics*, **54**, 61.

11. Brown, T. A. (ed.) (1998). *Molecular biology labfax*, 2nd edn, Vol. 1. Academic Press, London.

12. Hatt, H. (ed.) (1980). *Laboratory manual on freezing and freeze-drying.* American Type Culture Collection, Rockville.

13. Kirsop, B. E. and Snell, J. J. S. (1991). *Maintenance of microorganisms and cultured cells: a manual of laboratory methods*, 2nd edn. Academic Press, London.

14. Ptashne, M. (1992). *A genetic switch: phage λ and higher organisms*, 2nd edn. Blackwell, Oxford.

15. Miller, H. (1987). In *Methods in enzymology* (ed. S. L. Berger and A. R. Kimmel), Vol. 152, p. 145. Academic Press, New York.

Chapter 3

Purification of DNA

Paul Towner

College of Health Science, University of Sharjah, PO Box 27272, UAE

1 Introduction

This chapter explains how to purify genomic DNA and the DNA of different types of vectors. Genomic DNA can be obtained from any microorganism, plant or animal at any time during development and provides DNA which contains a copy of every gene from the organism. Vector DNA refers to plasmid and phage DNA.

The preparation of DNA from virtually every possible source has recently become much simpler because of the introduction of methods using the chaotropic reagent guanidinium thiocyanate, which is able to dissolve tissues and leave nucleic acids intact. After mixing this solution with a matrix such as diatomaceous earth, or filtering through glass fibre micropore filters, the DNA binds to the silica and can be recovered unsheared and in good yield. The basis of this technique was described some time ago (1) and subsequently used in the purification of DNA from many sources, including the preparation of plasmids (2). The DNA and RNA purification kits supplied by many commercial suppliers are founded on the initial observations described in these publications. The DNA prepared by this method can be used for PCR, Southern hybridization, construction of gene libraries, transfections and sequencing. Consequently, the previous methods used to prepare DNA have become obsolete and will not be described here.

2 Silica-based methods for the isolation of DNA

2.1 Description of the methodology

There are essentially four steps in the preparation of DNA using silica-based methods:

(1) Preparation of the sample of material from which the DNA is to be isolated

(2) Dissolution of the sample in the guanidinium solubilization buffer

(3) Application of the solubilized sample to the silica-based adsorbent

(4) Removal of the guanidinium buffer from the silica and elution of the DNA sample

The simplest materials to process are bacteria, yeast and tissue culture cells because they can easily be harvested by centrifugation and resuspended in water or isotonic saline and then mixed with the guanidinium solubilization buffer to form a transparent solution. Adherent tissue culture cells can be dissolved directly by pipetting on the solubilization buffer after aspiration of media. The DNA from viral particles such as λ and adenovirus can also be obtained very easily although the viral samples should be very pure and prepared by CsCl density gradient centrifugation (see Chapter 2, *Protocol 13*) to avoid the possibility of contamination with host DNA. Blood samples should be processed to remove contaminating erythrocytes. This can be achieved by layering the blood or bone marrow sample onto a 5.7% (w/v) Ficoll solution and centrifuging the sample at 4000 g for 45 min, which causes the erythrocytes to migrate to the bottom of the tube leaving the leucocytes at the interface. These can be removed with a Pasteur pipette, diluted with isotonic buffer and collected by centrifugation. Alternatively, the blood sample can be diluted with an equal volume of erythrocyte lysis buffer (155 mM NH_4Cl, 0.1 mM EDTA, 10 mM $NaHCO_3$) and left at room temperature for 30 min: the leucocytes can be recovered by centrifugation at 3300 g for 20 min. Samples from plants and animal tissues are difficult to obtain in a condition where they will readily dissolve in the guanidinium solubilization buffer. It is preferable to process them by freezing the sample in liquid nitrogen and grinding to a powder using a pre-cooled pestle and mortar. The frozen powder can then be sprinkled onto the solubilization buffer in a Petri dish or other container offering a large surface area. By dispersing the powder coagulation is avoided and the sample dissolves, giving good yields of DNA even with partly insoluble material such as bone and fibrous plant components. The purification of plasmid DNA from bacterial cultures is also very straightforward, inexpensive and gives a product which can be used for sequencing and transfections.

Several companies market proprietary kits for the isolation and recovery of genomic and plasmid DNA, the costs of these are becoming quite favourable because of the increased demand for molecular biological techniques in many laboratories. Most kits are based on the chaotropic properties of guanidinium thiocyanate which denatures and dissolves all components other than nucleic acid. Under these conditions, the DNA has a high affinity to bind to silica derivatives either in the form of diatomaceous earth suspended in the guanidinium solution or as glass-fibre filters. The concentration of the guanidinium thiocyanate for the process to be effective is not absolutely crucial. The stock solubilization buffer contains 4 M guanidinium but becomes diluted by the addition of resuspended samples to as low as 2.5 M; the yield and quality of the resulting DNA are apparently unaffected within this concentration range. The processing of the DNA–silica matrix can be performed using centrifuge tubes but requires extensive pipetting to remove supernatants between centrifugation steps. This makes the procedure time-consuming and is not as effective as the use of flow-through columns, which can be attached to a vacuum line. Columns of this type can be crudely prepared by piercing the base of a 0.6 ml microfuge tube with a 22G syringe needle. The columns are only effective when used with glass-fibre filter paper because diatomaceous earth frequently leaks out even when the hole is well packed with siliconized glass wool. It is preferable to use commercial systems which comprise a microfuge tube and a filter unit which fits inside it. These units are designed for centrifugation but the filter unit can be fitted directly into a rubber or polyethylene vacuum line attached to a water-jet pump. The filter units can be recycled if properly cleaned. Two types are available. The first contains a thin membrane, supported by a nylon mesh, fixed into the base of the tube. These units are available from Whatman and are also supplied in Hybaid kits. If used with diatomaceous earth their pores eventually become blocked, which makes them very difficult to clean without damaging the membrane. However, if a disc of a glass microfibre filter is placed on the membrane before use they are less prone to damage. A more robust filter unit is supplied with Qiagen kits and is composed of a tube into which can be slid a large-pore sintered disc, which is held in place by a nylon O-ring. These tubes can be cleaned by sonication to dislodge particles in the sinter, soaked in 10% nitric acid to oxidize any traces of nucleic acid, and then washed extensively before reassembly.

Glass microfibre filters can be prepared cost-effectively from large diameter sheets of Whatman GF/C or GF/D filters by punching out 7 mm diameter discs using a sharp cork borer. These discs, which are as useful as diatomaceous earth, can be placed between the sinter disc and nylon retaining ring. If diatomaceous earth is used the glass-fibre filter is not absolutely required but its presence will prevent the sinters from becoming blocked by small particles of silica. Large-scale preparations of DNA are most commonly used in plasmid preparations and require larger filter units which are also available from Whatman (VectaSpin columns) and can be fitted with glass microfibre filters of 22 mm diameter prepared using an appropriate cork borer.

2.2 Purification of genomic and viral DNA

Protocol 1 describes a typical methodology for solubilization of cell and tissue samples. It can be used with animal and plant tissues and cell cultures, with yeast, fungal and bacterial cells, and also with preparations of viruses such as phage λ. In *Protocol 2* DNA is purified from the solubilized material with diatomaceous earth and in *Protocol 3* purification is carried out with glass-fibre filters. Whichever method is used, the resulting DNA should be high molecular weight and suitable for any molecular biology application.

Protocol 1

Solubilization of cell and tissue samples

Reagents

• Guanidinium thiocyanate (Sigma G9277)

A. Preparation of the solubilization buffer

1 Dissolve 472 g guanidinium thiocyanate*ᵃ* in about 500 ml of water.

Protocol 1 continued

2 Add 50 ml 1 M Tris-HCl, pH 7.5, and 40 ml 0.5 M EDTA, pH 8.0, and make up to 1 l.

3 Store in an airtight screw-topped light-proof container.

B. Solubilization of cell and tissue samples

1 Prepare and harvest the sample to be processed and resuspend in a small volume of buffer or media to prevent lysis. Generally, a 50 μl pellet of bacteria, yeast, tissue culture cells or leucocytes or 20 μl viral samples (e.g. a phage λ stock prepared as described in Chapter 2, *Protocol 13*) can be resuspended in a 250 μl solution. Animal and plant tissues should be frozen and ground to a powder in liquid nitrogen using a pre-cooled pestle and mortar.

2 Add 1 ml of solubilization buffer to each 250-μl aliquot of resuspended material and mix by inversion to form a transparent solution. If the sample is slightly viscous or not entirely dissolved add further 200-μl increments of solubilization buffer. Tissues which have been powdered should be left at room temperature for a few minutes to allow the excess liquid nitrogen to evaporate and 100 mg samples sprinkled onto 1 ml of solubilization buffer.

[a] The loose hygroscopic crystals of guanidinium thiocyanate occupy around 600 ml. They dissolve endothermically, so it is expedient to use hot water.

Protocol 2

Purification of DNA using DNA-binding resin

Equipment and reagents

- Diatomaceous earth (Sigma D5384)
- Solubilization buffer (see *Protocol 1A*)
- Filter units (Whatman, Hybaid or Qiagen)

- Ethanol wash solution (50 mM Tris-HCl, pH 7.5, 10 mM EDTA, pH 8.0, 0.2 M NaCl, 50% [v/v] ethanol)

A. Preparation of the DNA-binding resin

1 Place 10 g of diatomaceous earth in a 2-l transparent vessel and stir in 2 l of water to resuspend the powder. Allow to settle for 1 h and pour away the supernatant containing any material which remains in suspension.

2 Resuspend the diatomaceous earth in a further 2 l of water and again allow to settle for 1 h before pouring away the supernatant.

3 Prepare solubilization buffer as in *Protocol 1A* but restrict the volume to 900 ml and use this to resuspend the diatomaceous earth before making up to 1 l.

4 Store in an airtight screw-topped light-proof container.

B. Purification of DNA

1 Resuspend the DNA-binding resin and pipette 0.5 ml into the solubilized samples from *Protocol 1*.

Protocol 2 continued

2 Mix very gently by inversion to avoid shearing and allow the DNA to bind to the resin.

3 Transfer the tube contents to a filter unit attached to a vacuum line and allow the liquid to drain from the filter.

4 Pipette 0.5 ml solubilization buffer into the filter unit to rinse the binding resin.

5 Pipette 2 ml ethanol wash solution into the tube, ensuring that all traces of liquid previously in the tube are displaced.

6 Transfer the filter unit into a microfuge tube and centrifuge for 2 min at 15 000 g to remove all traces of liquid.

7 Transfer the filter unit to a fresh microfuge tube then carefully pipette 20–50 µl water onto the bed of resin and allow to soak in.

8 Centrifuge for 2 min at 15 000 g to collect the liquid containing the DNA.

Notes

(a) The initial steps of *Protocol 2A* are designed to remove very small particles of silica which would otherwise block the holes in membranes and sintered discs. Consequently, the concentration of resin is actually less than 10 g/l, but the concentration is not critical. Stocks of up to 20 g/l can be prepared.

(b) The amount of DNA-binding resin required depends on the amount of DNA in the sample. One millilitre of a 10 g/l solution can bind in excess of 5 µg DNA. If there is a vast excess of nucleic acid the resin tends to clump very quickly and has a crystalline appearance; if this happens, then add more resin until the clumps disappear.

(c) The filter unit must be centrifuged to remove all traces of salt and ethanol because in most cases the DNA will be eluted in a very small volume of water. Have all reagents to hand so that air is not allowed to be sucked through the resin bed between the different wash steps. If traces of resin have leaked through the sinter during centrifugation then it is important to transfer the DNA solution to a fresh tube, avoiding any traces of resin. If filter units are not employed, all of the steps can be performed using a microfuge tube and the supernatants discarded by aspiration after brief centrifugation (10–20 s). Each of the solvent exchange steps should be duplicated to ensure that there is no carry-over of unwanted solutes.

Protocol 3

Purification of DNA using glass-fibre filters

Equipment and reagents

- Glass fibre filter discs, 7 mm diameter (Whatman)
- Solubilization buffer (see *Protocol 1A*)
- Ethanol wash solution (see *Protocol 2*)

Method

1 Prepare a filter column with five to ten 7-mm diameter glass-fibre filter discs. Attach to a water-jet pump at a very low flow rate.

Protocol 3 continued

2 Pipette the solubilized sample (see *Protocol 1*) onto the filter column such that 1 ml of the solution requires 1 min to flow through the column.

3 Pipette 0.5 ml solubilization buffer into the filter column to rinse the glass-fibre filter discs.

4 Pipette 2 ml ethanol wash solution into the filter column to remove all traces of solubilization buffer.

5 Transfer the filter column to a microfuge tube and centrifuge at 15 000 g for 2 min.

6 Place the filter column in a fresh microfuge tube and pipette 20–50 μl water onto the centre of the glass-fibre filter and allow it to soak in.

7 Centrifuge for 2 min at 15 000 g to collect the liquid containing the DNA.

Notes

(a) Unlike the use of diatomaceous earth, the DNA only has the opportunity to bind to the glass-fibre filter as it flows through the column. If the flow rate of the liquid is too high, substantial losses of DNA can result. After the initial binding step the water-jet flow rate can be increased.

(b) In the absence of filter units, glass-fibre filters can be packed into the base of a 0.6-ml microfuge tube which has been pierced at its apex.

2.3 Plasmid DNA

2.3.1 Small-scale 'mini-preps'

It is quite straightforward to isolate plasmid DNA from *Escherichia coli* on a 'mini-prep' scale starting from a few millilitres of culture and this can easily be scaled up to process much larger volumes (100 ml to 2 l). The yield and quality of the plasmid is very dependent on the conditions in which the bacterial culture is grown. A well-aerated culture flask with either SB or TB media, in preference to LB broth (see *Appendix 2*), will result in sufficient bacterial growth that after 4–6 h the sample can be processed. This is especially important when preparing very large plasmids, of 35 kb or more, because these are prone to deletion of non-essential sequences and readily form minicircle plasmids when growth conditions are poor. It is also important not to overgrow the culture because it then consists mostly of lysed bacteria and results in a poor-quality, low-yield preparation.

The key to successful isolation of plasmid DNA is the separation of the plasmids from the bacterial genome so that the resulting preparation is uncontaminated with bacterial DNA. The alkaline denaturation method (3) described in *Protocol 4* works as follows:

(1) The culture is harvested by centrifugation and resuspended in buffer

(2) The cells are mixed with lysis buffer containing NaOH and SDS; this raises the pH to 12.0–12.5 and causes the non-supercoiled genomic DNA to denature but has no effect on the supercoiled plasmid DNA

(3) The solution is mixed with a potassium acetate buffer to neutralize the alkali and precipitate the detergent as potassium dodecyl sulphate, with which the denatured genomic DNA coprecipitates

(4) The cleared supernatant containing the soluble plasmid DNA is recovered by filtration or centrifugation

(5) The plasmid DNA sample is purified using a guanidinium–silica matrix

Protocol 4

Small-scale isolation of plasmid DNA

Reagents

- SB or TB media (for recipes see *Appendix 2*)
- GTE buffer (25 mM Tris-HCl, pH 7.5, 10 mM EDTA, 50 mM glucose)
- Lysis buffer (0.2 M NaOH, 1% [w/v] SDS; prepared fresh from stocks of 5 M NaOH and 10% [w/v] SDS)
- Potassium acetate solution, prepared by adding glacial acetic acid to 5 M potassium acetate until pH 4.8 is obtained
- DNA-binding resin (see *Protocol 2A*) or solubilization buffer (see *Protocol 1A*)

Method

1   Prepare a culture of *E. coli* containing the plasmid in 3 ml of SB or TB media with sufficient aeration and the appropriate antibiotic. Colonies of bacteria on agar plates from an overnight incubation at 37 °C will have grown to a suitable density within 4–6 h if inoculated into pre-warmed media.

2   Transfer 1.5 ml of the culture to a microfuge tube and centrifuge at 15 000 *g* for 20 s to form a pellet.

3   Aspirate, discard the supernatant, and resuspend the cell pellet in 0.2 ml GTE buffer, ensuring that a homogeneous suspension results, with no cell clumps.

4   Add 0.2 ml lysis buffer and invert the capped tube several times to mix the contents. The solution should be translucent and viscous.

5   Add 0.2 ml potassium acetate solution and mix gently by inversion. The solution will lose its viscosity and a coagulated white mass of potassium dodecyl sulphate and bacterial genomic DNA will form.

6   Centrifuge at 15 000 *g* for 5 min to consolidate the precipitate which may either produce a pellet, pellicle or both.

7   Transfer the supernatant, about 0.6 ml, to a microfuge tube, add 1 ml of DNA-binding resin and process as described in *Protocol 2B*, or add 1 ml of solubilization buffer and purify DNA using glass-fibre filters as described in *Protocol 3*.

Notes

(a) It is not necessary to restrict the processed culture volume to 1.5 ml: all of the sample up to 5 ml can be used.

(b) Occasionally, RNA might contaminate the plasmid preparation. To ensure its removal, 2 μl of 10 mg/ml RNase A (See *Appendix 2*) can be added to the GTE buffer. After lysis (step 4) the solution is left at room temperature for 10 min to allow digestion of the RNA.

2.3.2 Large-scale preparation of plasmid DNA

Before proceeding with a large-scale culture of plasmid DNA it is worth pointing out that the amount of plasmid resulting from a small-scale preparation is usually enough for several restriction analyses and DNA-sequencing protocols. It is also very straightforward to process 50 ml cultures by scaling up the volumes used in *Protocol 4* tenfold. However, if a large stock of a plasmid is required for repetitive use then litre quantities of bacterial cultures will be required. The main problem that this introduces is that a large volume of cleared bacterial lysate is produced which requires substantial quantities of the guanidinium buffer. This requirement can be decreased by first precipitating the nucleic acid in the lysate with isopropanol and then digesting the RNA in the sample before mixing with the guanidinium–silica adsorbent. Depending on the plasmid–strain combination yields of plasmid in the range 4–10 mg/l of culture can routinely be obtained.

Protocol 5

Large-scale plasmid preparation

Equipment and reagents

- LB, YT or DYT broth (see *Appendix 2*)
- GTE buffer (see *Protocol 4*)
- Lysis buffer (see *Protocol 4*)
- Potassium acetate solution (see *Protocol 4*)
- DNA-binding resin (see *Protocol 2A*) (optional)
- Solubilization buffer (see *Protocol 1A*)

- Glass fibre filter discs, 7 mm diameter (Whatman)
- VectaSpin column (Whatman)
- Ethanol wash solution (see *Protocol 2*)
- TE buffer, pH 8.0 (10 mM Tris-HCl, pH 8.0, 1 mM EDTA, pH 8.0)

Method

1 Prepare a 1 l culture of *E. coli* harbouring the plasmid in a nutrient medium such as LB, YT or DYT, using sufficient aeration and an appropriate antibiotic. Conical flasks of 2.5 l capacity should contain at most 0.4 l of media. The media should be pre-warmed and inoculated using 10 ml of a fresh culture of bacteria and incubated for not more than 8 h.

2 Transfer the culture to large volume centrifuge buckets and centrifuge at 4000 g for 20 min.

3 Remove and discard the culture supernatant and resuspend the cell pellet in 50 ml GTE buffer, ensuring that a homogeneous suspension results, with no cell clumps.

4 Pour 50 ml lysis buffer into the cell suspension and gently swirl the mixture until a homogeneous translucent viscous gel is formed.

5 Pour in 50 ml of potassium acetate solution and mix gently by swirling or inversion to produce a clear supernatant and coagulated white mass.

6 Centrifuge at 5000 g or higher for 30 min to consolidate the white precipitate. If the supernatant remains slightly turbid it should be recentrifuged in 50 ml tubes to precipitate any particulate material.

7 Carefully remove the particle-free, clear supernatant and mix thoroughly with 250 ml of DNA-binding resin or solubilization buffer. If resin is used in the preparation, very gently agitate the mixture so that the resin is in suspension for several minutes.

8 If using resin, place one 22-mm diameter glass-fibre filter into a VectaSpin 20 column. If resin is not used then use 20 filters. Place a rubber collar around the column and insert it into the mouth of a side-arm flask connected to a suitable vacuum line.

9 If solubilization buffer is being used for the preparation, transfer the solution onto the stack of glass-fibre filters and allow the liquid to be drawn into the flask. If resin is being used, allow it to settle first so that a much smaller volume of suspension needs to be transferred to the column. Remove the vacuum line and transfer the filtrate to another vessel.

10 Reattach the vacuum and wash the column, containing either the stack of glass-fibre filters or resin, with 5 ml solubilization buffer followed by 20 ml of ethanol wash solution.

11 Remove the column from the flask, place in a 50 ml Falcon tube and centrifuge at 5000 g for 10 min to remove all trace of liquid from the column contents.

12 Place the column in a clean tube and carefully add 5 ml TE, pH 8.0, or water onto the surface of the silica and allow it to soak through. Centrifuge at 5000 g for 10 min to retrieve the sample of DNA.

Notes

(a) For removal of RNA, mix the cleared lysate resulting from step 6 with an equal volume of isopropanol and leave on ice for 20 min to precipitate the nucleic acid. After centrifugation (5000 g for 5 min), resuspend the precipitate in 10 ml 70% (v/v) ethanol, recentrifuge and partly dry the nucleic acid pellet to remove traces of ethanol. Redissolve the pellet in 20 ml of TE pH 8.0 containing 10 μg/ml RNase A (see *Appendix 2*) and incubate for 20 min. This solution can then be mixed with 40 ml of DNA-binding resin or solubilization buffer and processed. The removal of RNA from the sample allows a large reduction in the amount of silica required to bind the DNA.

(b) At step 9 it is wise to retain the filtrate because it can be recycled to ensure all the nucleic acid absorbs to the glass-fibre filters, or mixed with additional resin to increase yield. If VectaSpin columns are not available then ordinary chromatography columns with a suitable sinter may be used with either a vacuum line or positive pressure to pump the liquids through the matrix. The DNA will be contaminated with ethanol wash solution when it is eluted, so will need to be precipitated using ethanol prior to use (see *Protocol 8*).

2.4 Post-purification treatment of samples with phenol

The use of guanidinium and silica is generally sufficient for the preparation of pure DNA from very many sources of material, although occasionally DNA samples are obtained which appear pure by several criteria but do not behave as anticipated. It is then necessary to use aqueous phenol or phenol–chloroform as an aid to removal of substances which have co-purified with the DNA sample. Phenol is a very good solvent of proteins but can also dissolve small quantities of DNA. For this reason, it is preferable to use phenol–chloroform on DNA samples which are not heavily contaminated because DNA does not dissolve in this organic mixture. Following the use of these solvents the DNA sample should be washed with 24:1 (v/v) chloroform–isoamyl alcohol to remove traces of phenol prior to precipitation of the DNA using ethanol. *Protocol 6* explains how these various solvent mixtures are prepared, *Protocol 7* describes the phenol extraction procedure, and *Protocol 8* details ethanol precipitation.

Protocol 6

Preparation of phenol and chloroform solutions

Reagents

- Water-equilibrated phenol
- 1 M Tris-HCl, pH 8.0
- 50 mM Tris-HCl, pH 8.0
- Chloroform
- Isoamyl alcohol (the more correct name is 3-methyl-1-butanol)
- TE buffer, pH 8.0 (see *Protocol 5*)

A. Preparation of phenol

Caution: Phenol is one of the most hazardous chemicals used in molecular biology (see Chapter 1, Section 3.3) and should be obtained ready for use either as an aqueous solution or buffered at pH 8.0. Aqueous phenol must be pretreated by buffering to pH 8.0

Method

1. Remove and discard the upper aqueous layer from a 1-l bottle of water-equilibrated phenol. Replace the aqueous layer with 1 M Tris-HCl, pH 8.0, and mix the liquid phases by inverting the bottle several times.

2. Allow the layers to separate. If the lower layer is at pH 8.0 then remove and discard the upper aqueous phase. Replace this with 50 mM Tris-HCl, pH 8.0. Agitate the mixture and aliquot 1–10 ml samples in microfuge or Falcon tubes and store at −20 °C. If the lower layer is not at pH 8.0 then repeat step 1 until this pH is reached.

3. Thaw aliquots at 20 °C prior to use. It is the dense lower organic layer which is used for treatment of DNA. The phenol remains stable at 4 °C for about 1 month, but storage at room temperature or exposure to light enhances oxidation and the liquid becomes pink. Such samples must be discarded because they can lead to the modification and cleavage of DNA in solution. Oxidation can be inhibited by the addition of 8-hydroxyquinoline at 0.1% (w/v). This has the advantage of giving a bright yellow colour which identifies the organic layer during phenol extractions.

B. 24:1 (v/v) chloroform–isoamyl alcohol

1   Mix 96 ml chloroform with 4 ml isoamyl alcohol and store in a capped bottle.

C. 'Phenol–chloroform'

1   Thaw a 10 ml aliquot of phenol and transfer 5 ml of the lower phase to a universal bottle.

2   Mix in 5 ml of 24:1 (v/v) chloroform–isoamyl alcohol and 10 ml TE, pH 8.0, and store at 4°C. Use the lower organic layer.

Protocol 7

Phenol extraction

Reagents

- Phenol-chloroform (see *Protocol 6C*)
- 24:1 (v/v) chloroform-isoamyl alcohol (see *Protocol 6B*)

Method

1   Add 0.5 volume phenol–chloroform to 1.0 volume of aqueous DNA solution in a microfuge tube and mix by inversion to form an emulsion.

2   Centrifuge the tube at 15 000 g for 2 min to separate the sample into two phases. Remove the upper aqueous phase to a clean tube immediately, being careful to avoid the interface.

3   Add a half-volume of chloroform–isoamyl alcohol to the DNA sample, mix by inversion, centrifuge at 15 000 g for 2 min and remove the upper aqueous phase to a clean tube.

Notes

(a) Oligonucleotides, plasmids and genomic DNA can all be treated with phenol.

(b) It is advisable with samples of genomic DNA to mix solutions carefully to avoid shearing.

Protocol 8

Precipitation of DNA with ethanol

Reagents

- 3 M potassium acetate solution pH 5.6, prepared by adding glacial acetic acid to 3 M potassium acetate until this pH is obtained
- Absolute ethanol
- 70% (v/v) ethanol
- TE buffer pH 8.0 (see *Protocol 5*)

Protocol 8 continued

Method

1   Place a microfuge tube containing the DNA solution on ice and add a 0.1 volume of 3 M sodium acetate solution pH 5.6, and mix.

2   Add 2.5 volumes of cold absolute ethanol ($-20\,°C$) and mix by inversion. Incubate at $-20\,°C$ for 30 min or $-70\,°C$ for 10 min.

3   Centrifuge at 15 000 g for 30 min. Remove the supernatant and replace with 0.5 ml 70% (v/v) ethanol, mix by inversion and centrifuge at 15 000 g for 5 min.

4   Pour away the supernatant and dry the mouth of the inverted tube with tissue. Place the tube in a horizontal position at room temperature and allow the traces of ethanol to evaporate. Redissolve the DNA in water or TE, pH 8.0.

Notes

(a) Ethanol used for this procedure should be stored and used from the freezer. Ethanol at room temperature will cause high-molecular-weight DNA to shear.

(b) If an oligonucleotide is being processed then it must be more than 12 nucleotides in length and 3.0 volumes of ethanol must be used in step 2.

(c) If the amount of DNA exceeds 10 µg then the precipitate may be visible after addition of absolute ethanol. In this case the initial centrifugation in step 3 can be for 2 min.

3 Assessment of quality

The most straightforward procedure to assess the quantity and quality of plasmid and genomic DNA is by electrophoresis in an agarose gel with known standards. The guanidinium–silica purification method described here results in genomic DNA larger than λ (49 kb), as estimated in 0.3% (w/v) agarose gels (see Chapter 5), and provides intact plasmids of up to 35 kb. The electrophoresed DNA should appear as discrete bands with no discernible trail in the direction of migration, which would indicate extensive shearing. For more precise quantification, a sample of the DNA may be diluted and analysed by spectrophotometry (*Protocol 9*).

Protocol 9

Spectrophotometric determination of DNA

Equipment and reagents

- UV-visible scanning spectrophotometer
- Quartz cuvettes
- TE buffer, pH 7.5 (10 mM Tris-HCl, pH 7.5, 1 mM EDTA, pH 8.0)

Method

1   Transfer 10–100 µl of the DNA to 900–990 µl TE, pH 7.5, in a 1 ml quartz cuvette.

2   Mix the contents thoroughly and record a spectrum from 250–320 nm. A smooth

Protocol 9 continued

peak should be observed with an absorption maximum at around 260 nm and no noticeable shoulder at 280 nm (the latter indicates contamination with protein). The 260–280 nm absorbance ratio should not be more than 1.9. If the ratio is less than 1.9, contamination with protein should be suspected.

3 Calculate the quantity of DNA using the guide that 1 ml of a solution with an A_{260} of 1.0 is equivalent to 50 μg of double-stranded or 35 μg of single-stranded DNA.

Acknowledgement

I am indebted to Dr Nicholas Lea for suggesting the use of glass fibre filter discs and demonstrating the effectiveness of these protocols.

References

1. Boom, R., Sol, C. J. A., Salimans, M. M. M., Jansen, C. L., Wertheim-van Dillen, P. M. E., and van der Noordaa, J. (1990). *J. Clin. Microbiol.* **28**, 495.
2. Carter, M. J. and Milton, I. D. (1993). *Nucl. Acids Res.* **21**, 1044.
3. Birnboim, H. C. and Doly, J. (1979). *Nucl. Acids Res.* **7**, 1513.

Chapter 4
Purification of RNA

Miles Wilkinson

Microbiology and Immunology Department and the Vollum Institute, Oregon Health Sciences Laboratory, Portland OR 97201, USA

Additional material provided by T. A. Brown

1 Introduction

The most important consideration in the preparation of RNA is to rapidly and efficiently inhibit the endogenous ribonucleases which are present in virtually all living cells. There are several classes of ribonucleases which have been well characterized in eukaryotic and prokaryotic organisms, including both endonucleases and exonucleases. Methods designed for the preparation of RNA depend on speed and/or efficiency of neutralization of ribonucleases for their success. Most methods stipulate the use of ribonuclease inhibitors such as RNasin, vanadyl ribonucleoside complexes, guanidinium hydrochloride, guanidinium isothiocyanate, heparin, and dextran sulphate, to name a few. Proteinase K is sometimes used to degrade ribonucleases, and organic solvents such as phenol and chloroform are used to remove ribonucleases by extraction procedures. Another important concern in the preparation of RNA is to avoid accidental introduction of trace amounts of ribonucleases from hands, glassware and solutions. Finally, it is desirable that the methods used for the preparation of RNA are relatively rapid and simple. Literally hundreds of procedures for RNA preparation exist, many of which are time-consuming and labour-intensive. The methods described here are relatively simple to perform and provide intact RNA suitable for cDNA cloning, RNA (Northern) blotting, or *in vitro* translation. Techniques for the preparation of total cellular RNA from animal and plant cells, lower eukaryotes and prokaryotic cells are provided. Several alternative methods are given which allow for the purification of cytoplasmic RNA, nuclear RNA, and poly(A)$^+$ RNA from eukaryotic cells. A rapid mini-prep method for the isolation of cytoplasmic RNA from small numbers of eukaryotic cells is also provided. Techniques to prepare polysomal RNA and immunoprecipitate specific polysomal mRNAs are described elsewhere (1–3).

2 Ribonuclease-free conditions

It is worth devoting time towards the task of making a 'ribonuclease-free environment' in part of your laboratory. This includes setting aside an area of the

- For RNA samples at low concentration ($<10\,\mu g/ml$) it may be necessary to centrifuge at $10\,000\,\boldsymbol{g}$ in sturdy tubes (e.g. 15 ml Falcon 2059 tubes or 40 ml 'oakridge' tubes)

Free nucleotides (e.g. radioactively labelled NTPs) are efficiently ($>90\%$) removed by precipitation in the presence of 2.5 M ammonium acetate (0.5 volume of 7.5 M ammonium acetate) and 2.0–2.5 volumes of ethanol. Precipitate at room temperature for 15 min, and centrifuge for 10–30 min, depending on the amount of RNA being precipitated (see above).

After centrifuging the sample, examine the white RNA pellet: it should be visible if more than 3 µg of RNA is present. Carefully remove the supernatant without disturbing the pellet, leaving 5–20 µl of the supernatant behind. Add 1 ml 80% (v/v) ethanol (for samples in microfuge tubes), centrifuge for 1 min, and carefully remove the supernatant. The 80% ethanol wash removes salt and trace amounts of organic solvents which may have been present in your RNA sample. Dry the RNA sample by air-drying for at least 15 min. Drying in a speed-vac is more rapid, but care must be taken so that the dry RNA pellet does not become airborne and vacates the tube.

RNA stored at $-70\,°C$ in the presence of ethanol and salt is stable for several years. Storage at $-20\,°C$ is appropriate for short-term storage, but is not recommended for periods of greater than 1 month. Because RNA in this form is a precipitate, it is important to mix the contents of the stock tube vigorously before taking an aliquot to centrifuge. For convenience, RNA can also be frozen at $-70\,°C$ in water alone. If the RNA sample is absolutely free of ribonucleases, it can be stored for years in this form without detectable degradation.

5 Preparation of total cellular RNA

Four methods for the preparation of total cellular RNA are described below. The first method, the guanidinium–CsCl method, depends on the potent chaotropic agent guanidinium isothiocyanate to both lyse the cells and rapidly inactivate ribonucleases (6, 7). Since RNA has a higher buoyant density than DNA and most proteins, it is purified from the lysate by ultracentrifugation through a dense CsCl cushion. The advantage of this method is that samples can be quickly lysed and then stored for several days before ultracentrifugation. For example, this method is useful for kinetic studies where samples must be taken at several different time points. Guanidinium isothiocyanate is such a powerful denaturant of ribonucleases that this method can be effectively used for preparation of RNA from ribonuclease-rich tissue such as pancreas (6). The disadvantage of this method is that it requires an overnight ultracentrifugation step. However, this allows for the purification of genomic DNA, in addition to RNA. The second and third techniques also depend on guanidinium isothiocyanate as a denaturing agent, but do not require ultracentrifugation over CsCl (8, 9). All three of these methods can be used for preparation of RNA from most mammalian cells; they can be used for lower eukaryotic cells and prokaryotic cells by substituting an appropriate wash buffer in the first step. The fourth protocol, the 'hot phenol method', is specifically for prokaryotic cells. This method depends on the organic solvent phenol to extract and purify prokaryotic RNA (10). The hot phenol method provides high yields of RNA within hours, without an ultra-centrifugation step. However, this method requires speed on the part of the worker to avoid RNA degradation.

Protocol 1

Guanidinium–CsCl method for preparation of total cellular RNA

Reagents

- Tris-saline (25 mM Tris, 0.13 M NaCl, 5 mM KCl). For a 1 l solution, add 3 g Tris-base, 8 g NaCl and 0.36 g KCl. Adjust the pH to 7.2–7.4

- Guanidinium lysis buffer (4 M guanidinium isothiocyanate, 30 mM sodium acetate, 1 M β-mercaptoethanol). Dissolve 47 g guanidinium isothiocyanate (Sigma G9277) in about 50 ml water by heating to about 60 °C. Add 1 ml 3 M sodium acetate, pH 5.2. Add water to 93 ml and filter through a 0.45–0.80 μm pore size membrane or Whatman filter No. 1. Add 7 ml β-mercaptoethanol after filtration

- CsCl–EDTA (5.7 M CsCl, 5 mM EDTA). Dissolve 96 g of CsCl in 80 ml water, add 1 ml 0.5 M EDTA, pH 7.0, bring the volume exactly to 100 ml with water, and filter through a 0.20–0.45 μm pore size filter

- RNA resuspension buffer (0.4% [w/v] SDS, 5 mM EDTA)

- 3 M sodium acetate, pH 5.2, prepared by adding glacial acetic acid to 3 M sodium acetate until this pH is obtained

Method

1 Prepare the cells or tissues as follows:

  (a) **Adherent cells**. Wash twice with ice-cold Tris-saline. Remove the cells either with 1 mM EDTA in Tris-saline, trypsin/EDTA in a physiological buffer, or with a rubber policeman. Pellet the cells by centrifugation, decant off the supernatant, blot most of the residual liquid on a paper towel, vigorously tap the bottom of the tube with a finger to resuspend the pellet, and put on ice. Alternatively, if a small number of dishes are harvested, it is feasible to lyse the cells (step 2) while still attached to the dish.

  (b) **Suspension cells**. Wash once with ice-cold Tris-saline by centrifugation, decant off the supernatant, blot most of the residual liquid on a paper towel, vigorously tap the bottom of the tube with a finger to resuspend the pellet, and put on ice.

  (c) **Tissues**. One method to pulverize tissues is to grind the tissue with a cold mortar and pestle in the presence of lysis buffer and either dry-ice pellets or liquid nitrogen (step 2). A better approach is to homogenize the tissue in the presence of lysis buffer with a high-speed tissue solubilizer such as an omnimixer or polytron.

Protocol 1 continued

2   Add guanidinium lysis buffer to the single cell suspension,[a] and quickly vortex or tap the tube vigorously to completely lyse the cells. If it is desired to also prepare high molecular weight genomic DNA from the cells, it is best to avoid excessive mechanical stress (e.g. vortexing). The volume of guanidinium lysis buffer depends both on the number of cells and the size of the ultracentrifuge tubes to be used. In general, add at least 10 volumes of lysis buffer to a tissue fragment or cell pellet, or about 1 ml of lysis buffer per 3×10^7 dispersed mammalian cells. The volume of the cell lysate should be two-thirds the volume of the ultracentrifuge tube used in step 4.

3   Proceed to the ultracentrifugation step immediately, or store the cell lysate at $-70\,°C$ or $-20\,°C$. Lysates can usually be stored for several weeks in this form without detectable deterioration of the RNA.

4   Rinse the ultracentrifuge tubes with DEPC-treated water to remove possible debris from the tube. Add CsCl–EDTA to the ultracentrifuge tube so that it takes up one-third the volume of the tube.

5   Layer the cell lysate on top of the CsCl. Add more lysis buffer, if required, to bring the volume to within 2 mm of the top of the tube, and to balance the buckets.

6   Spin in the ultracentrifuge for at least 12 h at $18\,°C$ (lower temperatures may precipitate the CsCl). For most ultracentrifuge rotors (e.g. Beckman SW50.1 and SW60), 35 000–40 000 r.p.m. is sufficient to pellet the RNA. The TLS-55 rotor used with table-top Beckman ultracentrifuges must be spun a 55 000 r.p.m. for at least 3 h to pellet the RNA (11).

7   After the centrifugation, examine the tube. An RNA pellet may not be visible unless there is more than 100 μg of RNA. A DNA band is typically observed in the lower third of the tube if more than 10^7 mammalian cells were used to prepare the lysate. The DNA can be purified by standard procedures (12). If only RNA is desired, aspirate the supernatant, being careful to remove all of the DNA. Leave the tube inverted so that the proteins and ribonucleases present in the tube will drain away from the RNA pellet.

8   Cut off the bottom 1 cm of the ultracentrifuge tube with a razor blade and quickly resuspend the RNA pellet (visible or not) in 100 μl water or RNA resuspension buffer. Transfer the RNA to a microfuge tube containing 0.4 ml 4:1 (v/v) chloroform–butan-1-ol.[b]

9   Add another 100 μl of water or buffer to the ultracentrifuge tube, resuspend the residual pellet and transfer to the same tube containing chloroform–butan-1-ol. Repeat this operation a total of four times so that the RNA pellet is resuspended in a total volume of 400 μl water or RNA resuspension buffer. Vortex briefly, centrifuge for 2 min. Take the aqueous phase (upper) and transfer to a second microfuge tube containing chloroform–butan-1-ol. Vortex, and repeat the centrifugation.

10 To determine the yield of RNA, take 10 μl of the upper phase and quantify the RNA using a spectrophotometer (see Section 3).

Protocol 1 continued

11 Ethanol precipitate the RNA (remaining upper phase) with 0.1 volume of 3 M sodium acetate pH 5.2 and 2.0–2.5 volumes of ethanol (see Section 4).

[a] In some cases, particularly for tissue lysates, it may be desirable to add 0.4 g solid CsCl per millilitre of lysis buffer. This usually restricts the migration of cellular debris during ultra-centrifugation to the upper half of the tube.

[b] For preparation of RNA from tissues, it is advisable to extract twice with 25:24:1 (v/v/v) phenol–chloroform–isoamyl alcohol prior to chloroform–butan-1-ol extraction in step 8.

Protocol 2 describes a modification of the method originally described by Cathala *et al.* (8). An important advantage of this method is that the steps prior to overnight LiCl precipitation can be done relatively rapidly. Hence, it is a useful method under circumstances where many RNA samples must be harvested, particularly if they are to be harvested at different times. The critical step in this method is the precipitation of RNA by LiCl. If the number of cells in the cell lysate is too few, RNA may not be efficiently precipitated. Conversely, if the density of cells is too great, contaminants such as DNA may be co-precipitated along with the RNA. A similar method is described by Chomczynski and Sacchi (9), where RNA is prepared in a period of hours by guanidinium isothiocyanate lysis and phenol–chloroform extraction (*Protocol 3*).

Protocol 2

Guanidinium–LiCl method for preparation of total cellular RNA[a]

Reagents

- GTEM lysis buffer (5 M guanidinium isothiocyanate, 50 mM Tris-HCl, pH 7.5, 10 mM EDTA, pH 7.5, 8% [v/v] β-mercapto-ethanol). Dissolve 59 g guanidinium isothiocyanate (Sigma G9277) into about 50 ml of water by heating to approximately 60°C. Add 5 ml of 1 M Tris-HCl, pH 7.5, and 2 ml of 0.5 M EDTA, pH 7.5, bring to a volume of 92 ml with water, filter, and

  store at 4°C. Add β-mercaptoethanol to a concentration of 8% (v/v) just before use

- 6 M LiCl

- PK buffer (50 mM Tris-HCl, pH 7.5, 5 mM EDTA, pH 7.5, 0.5% [w/v] SDS)

- Proteinase K (20 mg/ml stock made in water and stored at −70 °C)

- 3 M sodium acetate, pH 5.2 (see *Protocol 1*)

Method

1 Refer to *Protocol 1*, steps 1–2, for instructions on how to wash and lyse the cells (substitute the GTEM lysis buffer for the guanidinium lysis buffer). Use at least 7 ml of GTEM lysis buffer for every gram of tissue. For cultured eukaryotic cells, lyse about 3×10^6 cells/ml of lysis buffer (the actual value for a given cell type is empirical).

2 Shear the DNA with an 18-gauge needle or with a tissue solubilizer (e.g. polytron at medium speed).

3 For tissues, add an equal volume of 24:1 (v/v) chloroform–isoamyl alcohol, vortex for 15 s, centrifuge at 2500 *g* for 10 min, and transfer the upper aqueous phase to another tube.

4 Add 1.4 volumes of 6 M LiCl and precipitate for at least 15 h at 4°C.[b]

5 Centrifuge for 30 min at 10 000 *g* at 4°C. Aspirate as much of the supernatant as possible.

6 Add a volume of PK buffer which is one-half of the original lysis volume.

7 Add proteinase K to a final concentration of 200 μg/ml (add 10 μl/ml), resuspend the pellet, and incubate for 30 min at 45°C.[c]

8 Add NaCl to a final concentration of 0.3 M. Extract with an equal volume of phenol, followed by one or more extractions with 25:24:1 (v/v/v) phenol–chloroform–isoamyl alcohol until no protein is evident at the interface. Extract once with 24:1 (v/v) chloroform–isoamyl alcohol. For each extraction, centrifuge for 10 min at 2500 *g* in conical tubes.

9 Ethanol precipitate with 2.0–2.5 volumes of ethanol (Section 4).

10 Pellet the RNA by centrifugation, resuspend the RNA in 400 μl water, take at least 5 μl for quantification (see Section 3), transfer to a microfuge tube and ethanol precipitate the RNA with 0.1 volume of 3 M sodium acetate, pH 5.2, and 2.0–2.5 volumes of ethanol (see Section 4).

[a] The instructions are intended for processing moderate to large numbers of cells ($> 3 \times 10^6$ mammalian cells) in either 15 ml Falcon 2059 or 40 ml 'oakridge' tubes. A mini-prep version (for 1 ml of cell lysate or less) can be performed in microfuge tubes; for the extraction steps, centrifuge at full speed for 2 min.

[b] Samples can normally be stored in LiCl for several days without detectable RNA degradation.

[c] If there is difficulty resuspending the pellet, alternate proteinase K incubations with pipetting or vortexing.

Protocol 3

Guanidinium–phenol–chloroform method for preparation of total cellular RNA

Reagents

- GTSSC lysis buffer (5 M guanidinium isothiocyanate, 0.5% [w/v] Sarkosyl, 25 mM sodium citrate, 8% [v/v] β-mercapto-ethanol). Dissolve 59 g guanidinium isothiocyanate (Sigma G9277) into about 50 ml of water by heating to approximately 60°C. Add 5 ml of 10% (w/v) Sarkosyl (N-lauroylsarcosine) and 3.3 ml of 750 mM sodium citrate pH 7.0, bring to a volume of 92 ml with water, filter and store at 4°C. Add β-mercaptoethanol to a concentration of 8% (v/v) just before use

Protocol 3 continued

- 2 M sodium acetate pH 4.0, prepared by adding glacial acetic acid to 2 M sodium acetate until this pH is obtained

- Water-saturated phenol (see Chapter 3, *Protocol 6*)

Method

1 Follow steps 1–2 of *Protocol 2* to lyse the cells and shear the DNA, substituting GTSSC lysis buffer for GTEM buffer.

2 Add a 0.1 volume of 2 M sodium acetate pH 4.0 and 1 volume of water-saturated phenol.

3 Mix and then add 0.1 volume of 49:1 (v/v) chloroform–isoamyl alcohol.

4 Mix and incubate at 4 °C for 15 min.

5 Centrifuge for 20 min at 10 000 g at 4 °C. Transfer the supernatant to a new tube.

6 Add 1 volume of 100% isopropanol to the supernatant and precipitate for 30 min at −20 °C.

7 Centrifuge for 10 min at 10 000 g at 4 °C. Aspirate as much of the supernatant as possible.

8 Add to the pellet a volume of GTSSC lysis buffer which is one-half of the original lysis volume.

9 Add an equal amount of 100% isopropanol and repeat steps 6 and 7.

10 Complete the procedure by following steps 9 and 10 of *Protocol 2*.

Notes

(a) As with *Protocol 2*, the procedure can be used as a maxi-or mini-prep.

(b) Several companies market RNA purification kits based on this method. Most of these kits use a special solution that combines the GTSSC lysis buffer with phenol, so fewer steps are required to obtain the RNA. The kits also allow DNA and protein to be purified in parallel with RNA. Examples are TRI Reagent (Molecular Research Centre) and Tri-Pure (Roche Diagnostics).

(c) A modified version of this procedure has been described that uses bromochloropropane instead of chloroform. This solvent is less toxic and gives a better phase separation with less contamination of the RNA sample with DNA (13).

Protocol 4 is a procedure developed for obtaining RNA from prokaryotic cells (10). In this method, the bacteria are lysed in a hot SDS solution, followed by extraction with phenol. The bacterial DNA remains associated with the cell debris at the interface of the tube during extraction. This method is not appropriate for preparation of RNA from most eukaryotic cells since the DNA is less likely to be trapped at the interface of the tube during extraction. However, to prepare RNA from non-bacterial cells, the DNA can be sheared with an 18-gauge needle in step 3.

Protocol 4

Hot phenol method for preparation of total cellular RNA

Reagents

- Phenol equilibrated with 0.1 M sodium acetate, pH 5.2 (see Chapter 3, *Protocol* 6)
- 3 M sodium acetate, pH 5.2 (see *Protocol* 1)
- RNA lysis solution (0.15 M sucrose, 10 mM sodium acetate, pH 5.2, 1% [w/v] SDS)

Method

1 Add 4 ml equilibrated phenol to 40 ml 'oakridge tubes'. Warm these tubes to 65 °C in a water bath; also warm the RNA lysis solution to 65 °C.

2 Add the RNA lysis solution to a dispersed bacterial cell pellet. Use at least 10 volumes of RNA lysis solution to process 1 volume of cells. Quickly transfer the cell lysate to a tube containing hot phenol. **Caution: phenol is one of the most hazardous chemicals used in molecular biology** (see Chapter 1, Section 3.3).

3 Gently invert the tube several times and incubate the tube for 10 min at 65 °C. Invert the tube twice during the 10 min incubation period.

4 Centrifuge the tubes at 10 000 *g* for 5 min at 4 °C.

5 Transfer the upper phase to a new tube containing phenol, avoiding the white interface layer which contains protein and DNA. Gently invert the tube several times and incubate at 65 °C and centrifuge as before.

6 Transfer the upper phase to a tube containing 25:24:1 (v/v/v) phenol–chloroform–isoamyl alcohol, vortex briefly, and centrifuge as before.

7 Transfer the upper phase to a tube containing 24:1 (v/v) chloroform–isoamyl alcohol, vortex, and centrifuge as before.

8 Ethanol precipitate the RNA: transfer the upper phase to a tube containing 0.1 volume of 3 M sodium acetate, pH 5.2 (final concentration is 0.3 M), add 2.0–2.5 volumes of ethanol, and precipitate at −20 °C (see Section 4).

9 Pellet the RNA by centrifugation, resuspend the RNA in 400 μl water; take at least 5 μl for quantification (Section 3), transfer to a microfuge tube and ethanol precipitate the RNA (Section 4).

Note

Although this method provides high yields of RNA, in some cases the resulting RNA may not be entirely free of contaminating proteins. In this case, there are two options. After ethanol precipitation, the sample can be resuspended in 400 μl water and extracted once with 25:24:1 (v/v/v) phenol–chloroform–isoamyl alcohol and once with 24:1 (v/v) chloroform–isoamyl alcohol, followed by another ethanol precipitation. Alternatively, the pellet can be resuspended in PK buffer (see *Protocol* 2), and incubated with 200 μg/ml proteinase K for 30 min, followed by extraction, and ethanol precipitation (see Section 4).

6 Cytoplasmic and nuclear RNA

6.1 Cytoplasmic RNA

6.1.1 Cell lysis and extraction

The procedure described in *Protocol 5* can be used to prepare cytoplasmic RNA from either tissues or cell lines (7). A step which is critical for this method is the 'lysis step' where the cell membrane is disrupted, allowing the release of relatively intact nuclei. The method described here utilizes Nonidet P-40 under isotonic conditions (physiological salt concentrations) to achieve cell membrane rupture. This method works for many cell types. However, the conditions may require alteration to suit some cell lines and tissues. Cells that are not efficiently lysed will not release all of their cytoplasmic RNA, and thus a low RNA yield will result. However, sensitive cells have fragile nuclei which are disrupted by high concentrations of detergent, liberating DNA. The nuclear stain toluidine blue-O (Sigma) can be used to assess the state of the nuclei. Add 0.1 volume of a toluidine blue-O stock solution (1% dissolved in DMSO) to the nuclei and examine under a light microscope. The ratio of NDD buffer to Tris-saline can be altered to achieve appropriate lysis and release of nuclei. Alternatively, the concentration of Nonidet P-40 and DOC can be independently altered. In addition, some cell types (e.g. fibroblasts) may require Dounce homogenization and/or a 5–10-min incubation prior to centrifugation.

The number of cells processed by this procedure is critical. Because of the large amount of protein present in most cells, only moderate numbers of cells can be processed by phenol–chloroform extraction. Generally, the maximum number of cells that can be processed per tube is 5×10^7 tissue culture cells. If the cells possess a high nuclear to cytoplasmic volume ratio it may be possible to process 2×10^8 cells (e.g. quiescent lymphocytes); however, typically only 2×10^7 adherent fibroblasts can be processed per tube. One cannot scale up in a single tube; the procedure usually does not work well with large numbers of cells in large tubes.

Protocol 5

Isolation of cytoplasmic RNA

Reagents

- Tris-saline (see *Protocol 1*)
- NDD lysis buffer: to 90 ml of Tris-saline, add 10 ml 10% (v/v) Nonidet P-40, 0.5 g DOC and 10 mg dextran sulphate
- 3 M sodium acetate, pH 5.2 (see *Protocol 1*)

Method

1 Prepare the cells or tissues as follows:

   (a) **Adherent cells**. Wash twice with Tris-saline, remove the cells with either 1 mM EDTA in Tris-saline, trypsin/EDTA in a physiological buffer, or with a rubber policeman. Wash the cells twice with Tris-saline by centrifugation, decant off the supernatant and put the cells on ice.

(b) **Suspension cells**. Wash twice with ice-cold Tris-saline by centrifugation, pour off the residual supernatant and put the cells on ice.

(c) **Tissues**. Prepare tissues as a single-cell suspension. The method used depends on the tissue type. The use of proteolytic enzymes such as collagenase and/or Dounce homogenization is often required.

2   Resuspend up to 5×10^7 cells with 2.5 ml of ice-cold Tris-saline in a 15 ml polypropylene conical tube. Either use a pipette or resuspend the cells prior to adding the Tris-saline by sharply tapping the bottom of the tube with a finger.

3   Add 2.5 ml of NDD lysis buffer, invert the tube 10 times (or Dounce homogenize for some cell lines and tissues; see text), and centrifuge for 3 min at 2500 g at 4 °C. This step is the most important for achieving a good RNA preparation. It is imperative that this step be done as quickly as possible to decrease ribonuclease degradation. A centrifuge with an efficient braking system is desirable.

4   Carefully remove all but the last 0.1 ml of supernatant (cytosolic fraction) and quickly transfer to a 15 ml conical tube containing 5 ml of ice-cold 25:24:1 (v/v/v) phenol–chloroform–isoamyl alcohol. Add 0.25 ml of 20% (w/v) SDS (final concentration: 1%) and 150 μl of 5 M NaCl (0.3 M final concentration, including the NaCl contributed by the Tris-saline). Vortex the tube for 30 s and centrifuge for 10 min at 2500 g.[a]

5   Discard the nuclear pellet, or prepare nuclear RNA (see the text). If the nuclear pellet is 'loose' with viscous DNA emanating from it, see Section 6.1.1 for measures to prevent nuclear lysis.

6   Once the cytosolic fraction has completed its centrifugation, examine the sample. The upper phase should be relatively clear and the interface should contain a white proteinaceous precipitate. If the upper phase is extremely turbid, it is likely that too many cells were processed. The turbidity problem is normally alleviated by increasing the volumes used for extraction.

7   Transfer the upper aqueous phase (leaving behind the protein at the interface) to a fresh 15 ml polypropylene conical tube containing 5 ml of 25:24:1 (v/v/v) phenol–chloroform–isoamyl alcohol, vortex briefly, and centrifuge as before. Repeat this step until no more protein is visible at the interface.

8   Extract as above with 24:1 (v/v) chloroform–isoamyl alcohol.

9   Ethanol precipitate the RNA by transferring to a 15 ml polypropylene conical tube containing 10 ml of ethanol (2.0–2.5 volumes). Precipitate at −20 °C for at least 20 min (see Section 4).

10 Pellet the RNA by centrifugation at 2500 g for 20 min. Resuspend the RNA in 400 μl of water and quantify by spectrophotometry (see Section 3). If the RNA in solution is not clear, it may contain trace amounts of lipids and/or protein. In this case, extract with 25:24:1 (v/v/v) phenol–chloroform–isoamyl alcohol in a microfuge tube

(2 min centrifugation) until the interface is clean, then extract once with 24:1 (v/v) chloroform–isoamyl alcohol. Ethanol precipitate the RNA by the addition of 0.1 volume of 3 M sodium acetate pH 5.2 and 2.0–2.5 volumes of ethanol (see Section 4).

[a] Some protocols dictate adding urea and EDTA during extraction with phenol–chloroform (16). The use of these agents is suggested for cells with a high content of ribonucleases. Instead of adding SDS and NaCl, add an equal volume of a 2× extraction buffer (7 M urea, 450 mM NaCl, 10 mM EDTA, 10 mM Tris-HCl, pH 7.4, and 1% [w/v] SDS) which should be prepared fresh before use.

6.1.2 Ribonuclease degradation

Protocol 5 typically provides intact RNA suitable for Northern blots (14). The dextran sulphate present in the lysis buffer and the ionic detergent SDS added before extraction are ribonuclease inhibitors, albeit incomplete inhibitors. The cell lysis step is performed as quickly as possible at cold temperatures to decrease the possibility of RNA degradation. The extraction step depends on phenol and chloroform to act as denaturing agents which will inactivate endogenous ribonucleases. These precautions are intended to inhibit ribonuclease attack, but some cell types possess ribonuclease activity which is sufficiently high (e.g. pancreatic exocrine cells) that intact RNA cannot be prepared by this method. One alternative is to use the strong denaturing agent guanidinium isothiocyanate to prepare cytoplasmic RNA (see Section 6.1.4).

Another alternative is to use other ribonuclease inhibitors during the preparation of cytoplasmic RNA. For example, vanadyl ribonucleoside complexes (Life Technologies) inhibit many of the known ribonucleases, but do not interfere with most enzymatic reactions (15). Vanadyl ribonucleoside complexes should be used at a concentration of 10 mM during the cell lysis step.

Some protocols dictate adding urea and EDTA during extraction with phenol-chloroform (16). Urea is a strong denaturing agent and EDTA chelates Mg^{2+} ions required for the activity of some ribonucleases. The use of these agents is suggested for cells with a high content of ribonucleases (see the footnote to *Protocol 5*).

6.1.3 Rapid mini-prep method

Protocol 6 provides a simple and efficient method of purification of cytoplasmic RNA (17) from small numbers of eukaryotic cells (as few as 10^5–10^6 cells). Using this mini-prep method, RNA can be prepared from 10 different cell samples in considerably less than 2 h. The mini-prep method will allow purification from as many as 2×10^6 to 2×10^7 cells (depending on the cell type). The RNA yield from the mini-prep method is about 10 μg/10^6 cells, although this value can vary considerably depending on the cell type. The method is essentially a scaled-down version of *Protocol 5*. From most cell types, the quality of RNA isolated by this mini-prep method is suitable for Northern analysis or PCR (see Volume II, Chapters 5 and 7). If problems with RNA degradation or cell lysis are encountered, see Sections 6.1.1 and 6.1.2 for possible solutions.

Protocol 6

Rapid mini-prep of cytoplasmic RNA

Reagents

- Tris-saline (see *Protocol 1*)
- NDD lysis buffer (see *Protocol 5*)

Method

1 Pellet the cells, resuspend in 1 ml of ice-cold Tris-saline, transfer to a 1.5 ml microfuge tube and centrifuge at low speed (3000–6000 g) at 4°C for 30 s. If only a high-speed microfuge is available, centrifuge at 10 000 g for 10 s; however, this may seriously disrupt the cells.

2 Discard the supernatant and resuspend the cells in 250 μl ice-cold Tris-saline. Add 250 μl ice-cold NDD lysis buffer, invert the tube 10 times and centrifuge at 4°C for 30 s, preferably at low speed (3000–6000 g).

3 Carefully transfer the supernatant (being careful to avoid disturbing the nuclear pellet) to a microfuge tube containing 0.5 ml of 25:24:1 (v/v/v) phenol–chloroform–isoamyl alcohol. Quickly add 25 μl of 20% (w/v) SDS and 15 μl of 5 M NaCl to the tube, vortex or shake for 15 s, and microfuge at maximum speed at 4°C for 2 min. Discard the nuclear pellet or prepare nuclear RNA as described in Section 6.2.

4 Transfer the upper aqueous phase to another microfuge tube containing 25:24:1 (v/v/v) phenol–chloroform–isoamyl alcohol, vortex, and centrifuge at room temperature. Repeat this extraction procedure until there is no visible protein at the interface (normally two or three extractions are required). If the extraction is difficult because of excess turbidity or viscous DNA contamination see Section 6.1.1.

5 Extract once with 24:1 (v/v) chloroform–isoamyl alcohol to remove residual phenol.

6 Transfer the upper aqueous phase to a microfuge tube containing 1 ml of ethanol and precipitate at −20°C for at least 20 min (see Section 4).

6.1.4 Guanidinium method

In most circumstances, *Protocols 5* and *6* provide high yields of intact cytoplasmic RNA. However, for cell types with high levels of ribonuclease activity or transcripts which are highly unstable, these methods may not be appropriate. The following method provides high-quality cytoplasmic RNA from moderate numbers of cells.

Protocol 7

Guanidinium method for cytoplasmic RNA

Reagents

- Tris-saline (see *Protocol 1*)
- NDD lysis buffer (see *Protocol 5*)
- Guanidinium lysis buffer (see *Protocol 1*)

Protocol 7 continued

Method

1 Lyse the cells as described in *Protocol 6*, steps 1 and 2.

2 Transfer the cytoplasmic supernatant (0.5 ml) to a tube containing 2.5 ml of guanidinium lysis buffer.

3 Ultracentrifuge the sample over CsCl and process the RNA as described in *Protocol 1*, steps 2–11.

6.2 Nuclear RNA

Nuclear RNA is useful for studies examining precursor transcripts and RNA splicing intermediates. *Protocol 8* is an effective method for obtaining nuclear RNA (7). To determine whether the RNA that has been prepared is enriched for nuclear RNA, it is best to examine the sample by gel electrophoresis. One approach is to stain a suitable gel (Chapter 5, Section 4) with acridine orange to visualize the rRNA precursor transcripts. *Figure 1* shows that eukaryotic nuclear RNA prepared by this method is enriched for 32S and 45S rRNA precursor transcripts which are absent in cytoplasmic RNA. Both cytoplasmic and nuclear RNA contain mature 18S and 28S rRNA. These transcripts can also be observed directly on nylon-based transfers (see Volume II, Chapter 5) by staining with 0.03% (w/v) methylene blue and 0.3 M sodium acetate pH 5.2 for 45 s, followed by destaining with water for 1–2 min. The transfers must be baked or UV cross-linked prior to staining.

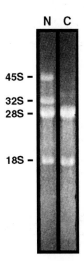

Figure 1 Ten micrograms of cytoplasmic (C) and nuclear (N) RNA from SL12.4 T lymphoma cells were prepared as described in *Protocols 5* and *8*, respectively, and electrophoresed in a 1% denaturing agarose gel containing formaldehyde (14). After electrophoresis, the gel was stained for 2 min in 10 μg/ml acridine orange, 10 mM sodium phosphate (1:1 ratio of monobasic and dibasic) and 1.1 M formaldehyde, followed by destaining for 20 min in the same buffer lacking acridine orange.

Protocol 8

Preparation of nuclear RNA

Reagents

- Guanidinium lysis buffer (see *Protocol 1*)

Method

1 Release the nuclei from large numbers of cells (> 1–5×10^7) as described in *Protocol 5*, steps 1–4. Smaller numbers of cells can be lysed as described in *Protocol 6*, steps 1–3.

2 Discard the excess supernatant from above the pellet with a Gilson P200 pipetter. Vigorously tap the bottom of the tube to resuspend the pellet. Perform this step rapidly so that the pellet does not warm to room temperature.

3 Add 1 ml of ice-cold guanidinium lysis buffer per 3×10^7 dispersed mammalian cells or add 10 volumes of lysis buffer compared with the cell pellet, and immediately agitate the tube so the pellet dissolves and forms a clear solution. If preparing DNA, do not vortex, instead vigorously tap the bottom of the tube with a finger to dissolve the pellet.

4 Add an appropriate amount of guanidinium lysis buffer to the nuclear lysate so that its final volume is equal to two-thirds of the volume of the ultracentrifuge tube to be used. Ultracentrifuge the sample through a CsCl cushion and process the RNA as described in *Protocol 1*, steps 2–11.

Note

An optional step prior to nuclei lysis is to wash the nuclei in ice-cold TNM buffer (0.1 M Tris-HCl, pH 7.5, 10 mM NaCl, 3 mM MgCl$_2$). Add TNM buffer carefully so the nuclear pellet does not become resuspended and centrifuge at 1000–2000 g for 3 min at 4°C. This wash step removes most of the residual cytoplasmic RNA present. However, the extra time required for this step could potentially lead to some degradation of the nuclear RNA.

7 Poly(A)$^+$ RNA

Ribosomal RNA is, by far, the most abundant RNA species in eukaryotic cells, representing 80–90% of the total cellular RNA. The two most abundant rRNA transcripts, 18S and 28S, have sizes of approximately 2 kb and 5 kb, respectively. In contrast, mRNA represents only 1–5% of the total cellular RNA, and is heterogeneous in size, ranging from less than 0.5 kb to over 10 kb. Most of the mRNA transcripts present in mammalian cells are polyadenylated at the 3'-end. The tail of adenine residues typically extends 50–200 bases. This poly(A) tail allows mRNA to be purified by affinity chromatography on oligo(dT)-cellulose. Described below are two methods which are based on oligo(dT)-cellulose chromatography. The first technique allows up to 10 mg of RNA to be processed per millilitre of oligo(dT)-cellulose (18). The principle of the method is that the RNA is bound in the presence of high concentrations of salt, washed in intermediate salt concentrations, and eluted in a low salt buffer. Yields are increased by recycling the RNA over the column twice. The second method is a rapid method of preparation of poly(A)$^+$ RNA directly from cell lysates (19). The advantage of this method is that poly(A)$^+$ RNA can be prepared from cells in about 4 h. A potential disadvantage to this approach is that it is not recommended for cells that possess a high content of ribonucleases.

Protocol 9

Enrichment for poly(A)$^+$ RNA from nuclear, cytoplasmic, or total RNA

Reagents

- Ribonuclease-free chromatography column (e.g. Bio-Rad Econo-Pac polypropylene column)
- Oligo(dT)-cellulose
- Regeneration solution (0.1 M NaOH, 5 mM EDTA)
- 1× loading buffer (50 mM sodium citrate, pH 7.5, 0.5 M LiCl, 1 mM EDTA, 0.1% [w/v] SDS)
- 2× loading buffer (100 mM sodium citrate, pH 7.5, 1 M LiCl, 2 mM EDTA, 0.2% [w/v] SDS)
- Wash buffer (50 mM sodium citrate, pH 7.5, 0.1 M LiCl, 1 mM EDTA, 0.1% [w/v] SDS)
- Elution buffer (10 mM sodium citrate, pH 7.5, 1 mM EDTA, 0.05% [w/v] SDS)
- 3 M sodium acetate pH 5.2 (see *Protocol 1*)

Method

1 Prepare a ribonuclease-free chromatography column by rinsing with freshly prepared 0.1% (v/v) DEPC in water, followed by autoclaving.

2 Suspend at least 25 mg of oligo(dT)-cellulose in regeneration solution for every mg of RNA to be processed. Transfer the oligo(dT)-cellulose to the column. The packed bed volume is 0.1 ml/25 mg of oligo(dT)-cellulose.

3 Wash the column with three column volumes of water. Wash with more water if the column effluent has a pH of greater than 8.0.

4 Equilibrate the oligo(dT)-cellulose with five column volumes of 1× loading buffer.

5 Dissolve the RNA in water to a final concentration of 5 mg/ml, heat to 65 °C for 5 min, and cool to room temperature. Warming the RNA helps to remove secondary structures which could interfere with poly(A)$^+$ RNA binding to the oligo(dT).

6 Add an equal volume of 2× loading buffer to the RNA, load on the column, wash with one column volume of 1× loading buffer, and collect the combined effluent.

7 Pass the RNA effluent over the column a second time to improve the yield: heat the effluent to 65 °C for 5 min, cool to room temperature, and load on the column.

8 Wash with ten column volumes of 1× loading buffer.

9 Wash with five column volumes of wash buffer.

Protocol 9 continued

10 *Optional:* measure the OD of fractions collected after the addition of wash buffer. If the final washes are devoid of measurable RNA, proceed to the next step. If significant amounts of RNA are being eluted, wash the column with more wash buffer.

11 Add three column volumes of elution buffer to elute the poly(A)$^+$ RNA. The RNA which is eluted is typically 20-fold enriched for poly(A)$^+$ RNA and is suitable for many applications, but still possesses over 50% rRNA contamination. If additional purification is required, proceed to the next step. Otherwise, ethanol precipitate the RNA as described in step 14.

12 Equilibrate the column with five volumes of 1× loading buffer. Incubate the eluate at 65°C for 5 min, allow to cool to room temperature, and repeat steps 6–11.

13 Take an aliquot of the eluted RNA for quantification (see Section 3). Typically, 1–3% of the input RNA is recovered after oligo(dT)-chromatography.

14 Ethanol precipitate the RNA by the addition of sodium acetate pH 5.2 to a final concentration of 0.3 M, and 2.0–2.5 volumes of ethanol (see Section 4).

15 Regenerate the column by sequential washing with three column volumes of regeneration solution, water and 1× loading buffer.

16 Store the oligo(dT)-cellulose at 4°C in 1× loading buffer containing 0.05% (w/v) sodium azide.

Protocol 10

Poly(A)$^+$ RNA prepared directly from cell lysates

Reagents

- Oligo(dT)-cellulose
- Poly(A)$^+$ lysis buffer (0.2 M NaCl, 0.2 M Tris-HCl, pH 7.5, 2% [w/v] SDS, 0.15 mM MgCl$_2$, 200 µg/ml proteinase K). To prepare 10 ml of buffer, add the following to 6.5 ml of water: 0.4 ml 5 M NaCl, 2 ml 1 M Tris-HCl, pH 7.5, 1 ml 20% (w/v) SDS, 1.5 µl 1 M MgCl$_2$ and add 100 µl of 20 mg/ml proteinase K immediately before use

- Ribonuclease-free chromatography column (see *Protocol 9*)
- Binding buffer (0.5 M NaCl, 10 mM Tris-HCl, pH 7.5)
- 3 M sodium acetate pH 5.2 (see *Protocol 1*)

Method

1 Prepare the cells or tissues as follows:

(a) **Cultured cells.** Wash as described in *Protocol 1*, step 1(a) or 1(b). Add 10 ml of lysis buffer per 10^8 cells and quickly homogenize the cell suspension in a high-speed tissue solubilizer (e.g. polytron) or by shearing the DNA through an 18-gauge needle until no longer viscous.

(b) **Tissue.** Quick freeze in liquid nitrogen. Grind up the tissue in a chilled mortar

Protocol 10 continued

and pestle, transfer the slurry to a 50-ml conical tube, and add 10 ml of lysis buffer per gram of tissue. Vortex until no tissue fragments are visible. Alternatively, pulverize the tissue in the presence of the lysis buffer with a high-speed tissue solubilizer.

2 Incubate the cell lysate for at least 1 h at 45°C. Either provide constant agitation in a shaking water bath, or agitate by hand every 10 min.

3 Hydrate the oligo(dT)-cellulose with water in a 50 ml conical tube. Prepare 0.2 g of oligo(dT)-cellulose per 10^9 cells or 10 g of tissue. Pellet by brief centrifugation and equilibrate the cellulose in 20-fold excess of binding buffer. Pellet and remove all but 1 ml of the binding buffer.

4 Adjust the NaCl concentration of the cell lysate to that of the binding buffer by adding 60 µl of 5 M NaCl per millilitre of lysate. Add the oligo(dT)-cellulose to the lysate, mix well, and incubate at room temperature for 30 min with constant agitation.

5 Pellet the oligo(dT) at room temperature (lower temperatures may precipitate the SDS).

6 Aspirate or pour off the supernatant and resuspend the pellet in an equal volume of binding buffer. Repeat until the supernatant appears clear.

7 Transfer the washed oligo(dT)-cellulose to a ribonuclease-free chromatography column and continue washing with three volumes of binding buffer or until the UV absorbance (260 nm) of the output gives a reading of less than 0.05 (see Section 3).[a]

8 Elute the poly(A)$^+$ RNA with three column volumes of water.

9 Precipitate the RNA by adding 0.1 volume 3 M sodium acetate pH 5.2 and 2.0–2.5 volumes of ethanol (see Section 4).

10 The yield of poly(A)$^+$ RNA from 10^8 cells ranges from 5–100 µg.

11 The oligo(dT)-cellulose can be regenerated as described in *Protocol 9*, step 15.

[a] If the column is running slowly due to the presence of high molecular weight genomic DNA, resuspend the contents of the column with a pipette.

Acknowledgements

I wish to thank Carol MacLeod, Patricia Salinas, Livia Theodor, Randy McCoy, and Jim Garret who introduced me to some of the methods described herein. I am also grateful to Thomas Herrick and Elisa Burgess who contributed data and provided ideas concerning these methods. I am indebted to Scott Landfear for his helpful hints and valuable discussion.

References

1. Schutz, G., Kieval, S., Groner, B., Sippel, A. E., Kurtz, D. T., and Feigelson, P. (1977). *Nucl. Acids Res.* **4**, 71.
2. Gough, N. M. and Adams, J. M. (1978). *Biochemistry* **17**, 5560.
3. Shapiro, S. Z. and Young, J. R. (1981). *J. Biol. Chem.* **256**, 1495.

4. Kumar, A. and Lindberg, U. (1972). *Proc. Natl. Acad. Sci. USA*, **69**, 681.

5. Zeugin, J. A. and Hartley, J. L. (1985). *Focus* **7(1)**, 1.

6. Chirgwin, J. M., Przybyla, A. E., MacDonald, R. J., and Rutter, W. J. (1979). *Biochemistry* **18**, 5294.

7. Wilkinson, M. (1988). *Nucl. Acids Res.* **16**, 10934.

8. Cathala, G., Savouret, J., Mendez, B. *et al.* (1983). *DNA* **2**, 329.

9. Chomczynski, P. and Sacchi, N. (1987). *Anal. Biochem.* **162**, 156.

10. VonGabain, A., Belasco, J. G., Schottel, J. L., Chang, A. C. Y., and Cohen, S. N. (1983). *Proc. Natl. Acad. Sci. USA* **80**, 653.

11. Verma, M. (1988). *Biotechniques* **6**, 848.

12. Iverson, P. L., Mata, J. E., and Hines, R. N. (1987). *Biotechniques* **5**, 521.

13. Chomczynski, P. and Mackey, K. (1995). *Anal. Biochem.* **225**, 163.

14. Wilkinson, M. and MacLeod, C. L. (1988). *EMBO J.* **7**, 101.

15. Berger, S. L. and Birkenmeier, C. S. (1979). *Biochemistry* **18**, 5143.

16. Pearse, M., Gallagher, P., Wilson, A. *et al.* (1988). *Proc. Natl. Acad. Sci. USA* **85**, 6082.

17. Wilkinson, M. (1988). *Nucl. Acids Res.* **16**, 10933.

18. Aviv, H. and Leder, P. (1972). *Proc. Natl. Acad. Sci. USA* **69**, 1408.

19. Badley, J. E., Bishop, G. A., St John, T., and Frelinger, J. A. (1988). *Biotechniques* **6**, 114.

Chapter 5

Nucleic acid electrophoresis in agarose gels

Douglas H. Robinson

Affymetrix Inc., 3380 Central Expressway, Santa Clara, CA 95051, USA

Gayle J. Lafleche

BioWhittaker Molecular Applications Inc., 191 Thomaston Street, Rockland, ME 04841, USA

1 Introduction

Electrophoresis is defined as the movement of ions and charged macromolecules through a medium when an electric current is applied. Agarose and polyacrylamide are the primary stabilizing media used in the electrophoresis of macromolecules.

Macromolecules are separated through the matrix based on size, charge distribution and structure. In general, nucleic acids migrate through a gel based on size, with little influence from base composition or sequence, whereas proteins separate through the matrix based on size, structure and charge because their charge density is not directly proportional to size.

Two equations are relevant to the use of power supplies for electrophoresis of macromolecules: Ohm's law and the second law of electrophoresis. These two laws and the interactions of these parameters (watts, volts, current) are critical to understanding electrophoresis.

1.1 Voltage, current and power: interactive effects on gel electrophoresis

Ohm's Law states that:

$$I = V/R$$

where I = current, V = voltage and R = resistance. This tells us that the current is directly proportional to the voltage and inversely proportional to the resistance. Resistance of the system is determined by the buffers used, the types and configurations of the gels being run, and the total volume of all the gels being run.

The second law:

$$W = I \times V$$

states that power or watts (a measure of the heat produced) is equal to the product of the current and voltage. Since $V = I \times R$, this can also be written as $W = I^2 \times R$.

During electrophoresis, one of the parameters is held constant and the other two are allowed to vary as the resistance of the electrophoretic system changes. In vertical systems, the resistance of the gel increases as highly conductive ions such as Cl^- are electrophoresed out of the gel. As these ions are removed from the gel, the current is carried by less conductive ions such as glycine, borate, acetate, etc. Under normal conditions in horizontal systems, there is little change in resistance. However, with high voltage or extended runs in horizontal systems, resistance can decrease.

1.1.1 Constant electrophoretic parameters

There are advantages and disadvantages for setting each of the critical parameters as the limiting factor in electrophoresis. Sequencing gels are often run at constant wattage to maintain a uniform temperature. Agarose and acrylamide gels for protein and DNA resolution are generally run at constant voltage. The relevant issues are as follows.

1.1.1.1 Constant wattage

In a vertical system when wattage is held constant, the velocity of the samples will decrease because the current, which is in part carried by the DNA, decreases to compensate for the increase in voltage. The generation of heat will remain uniform, resulting in a constant gel temperature. If the current should decrease disproportionately (from a buffer problem, buffer leak or a hardware problem), the power supply will increase the voltage to compensate. Since voltage and current vary over time at a constant wattage, it is not possible to predict the mobility of samples from the calculation of watt-hours.

1.1.1.2 Constant current

When the current is held constant, the samples migrate at a constant rate. Voltage and wattage will increase as the resistance increases, resulting in an increase in heat generation during the run. If a break occurs in the system, such as a damaged lead or electrode or a buffer leak, the resistance of the gel will be greatly increased. This will cause a large increase in wattage and voltage resulting in the generation of excessive heat. It is possible even for the system to get hot enough to boil, or scorch or burn the apparatus.

1.1.1.3 Constant voltage

When voltage is set constant, current and wattage will decrease as the resistance increases, resulting in a decrease of heat and DNA migration. Since the heat generated will decrease, the margin of safety will increase over the length of the run. If a problem develops and the resistance increases dramatically, the current and wattage will fall since the voltage cannot increase. Even if the apparatus fails, the worst that is likely to happen is that the resistance will increase so much that the power supply will not be able to compensate and it will shut off.

1.2 Physical chemistry of agarose

Agarose is a natural polysaccharide isolated from agar, which is obtained from various species of marine red algae. Agarose has become an important material in life science research and related areas because of its distinctive physical and chemical properties as a gel matrix.

1.2.1 Agarose as a gel matrix

Agarose is derived from a series of naturally occurring compounds present in seaweed. Most agar comes from various species of *Gelidium* and *Gracilaria*. All species contain ester sulphates and some, except *Gracilaria*, contain varying amounts of pyruvates. *Gracilaria* agarose contains methyl ethers, the position of which is variable according to the species.

Agarose consists of 1,3-linked β-D-galactopyranose and 1,4-linked 3,6-anhydro-α-L-galactopyranose. This basic agarobiose repeat unit (*Figure 1*) forms long chains with an average molecular mass of 120 kDa, representing about 400 agarobiose units (1). There are also charged groups present on the polysaccharide, most notably pyruvates and sulphates.

The advantages of agarose as a gel matrix are:

- Agarose forms a macroporous matrix which allows rapid diffusion of high molecular weight (1000 kDa range) macromolecules without significant restriction by the gel.

- Agarose gels have a high gel strength, allowing the use of concentrations of 1% or less, while retaining sieving and anticonvective properties.

- Agarose is non-toxic and, unlike polyacrylamide, contains no potentially damaging polymerization byproducts. There is no free radical polymerization involved in agarose gelation.

- Rapid staining and destaining can be performed with minimal background.

- Agarose gels are thermoreversible. Low-gelling- temperature and low-melting-temperature agaroses permit easy recovery of samples, including sensitive heat-labile materials.

- Agarose gels may be air dried.

1.2.2 Properties of agarose

The charged groups present on the polysaccharide—pyruvates and sulphates—are responsible for many of the properties of agarose. By careful selection of raw

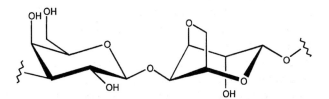

Figure 1 Agarobiose: the basic repeating unit of agarose.

materials, these properties can be controlled to meet specific needs. The details of these properties are described in the following sections.

1.2.2.1 Electroendosmosis

EEO is a functional measure of the number of sulphate and pyruvate residues present on the agarose polysaccharide (2, 3). This phenomenon occurs during electrophoresis when the anticonvective medium (the agarose in this case) has a fixed negative charge. In an electric field, the hydrated positive ions associated with the fixed anionic groups in the agarose gel migrate toward the cathode. Water is thus pulled along with the positive ions and migration of negative molecules such as DNA is retarded. Electroendosmosis is quantified by subjecting a mixture of dextran and albumin to electrophoresis, then visualizing them and measuring their respective distances from the origin. The amount of EEO $(-m_r)$ is calculated by dividing the migration distance of the neutral dextran (od) by the sum of the migration distances of the dextran and the albumin (od + oa):

$$-m_r = od/(od + oa).$$

1.2.2.2 Gelation

The mechanism for gelation of agarose was first suggested by Rees (4) and later demonstrated by Arnott *et al.* (5). It involves a shift from a random coil in solution to a double helix in the initial stages of gelation, and then to bundles of double helices in the final stage (*Figure 2*). The average pore size varies with concentration and type of agarose, but is typically 100–300 nm (6).

1.2.2.3 Melting and gelling temperatures

The energy needed to melt an agarose gel increases as the gel concentration increases (*Figure 3*). The gelling temperature of an agarose gel is also influenced

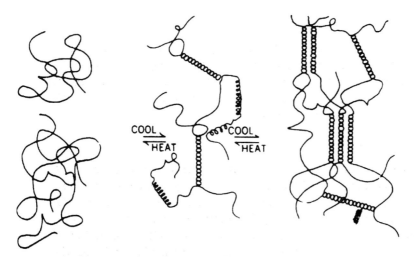

Figure 2 The gelation mechanism for agarose.

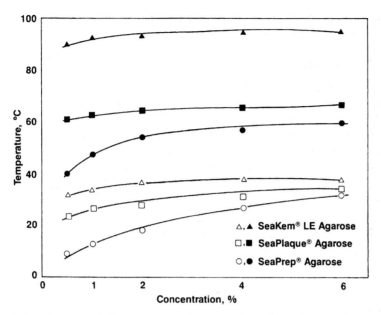

Figure 3 The effect of agarose concentration on the gelling and melting temperatures of agarose. The open symbols show the gelling temperature and the closed symbols show the melting temperature.

by the gel concentration. For this reason, gelling or remelting temperatures are expressed at a given agarose concentration. This property is of practical value since it is possible to vary gelling and melting parameters by using lower or higher concentrations of agarose. The dependence of gelling and melting temperatures on concentration is most pronounced at concentrations less than 1%.

1.2.2.4 Methylation of agarose

The agarose polysaccharide also contains uncharged methyl groups. The extent of natural methylation is directly proportional to the gelling temperature. Unexpectedly, synthetically methylated agaroses have lower, rather than higher gelling temperatures and the degree of synthetic methylation is inversely proportional to the melting temperature.

1.2.2.5 Gel strength

One of the most important factors contributing to the success of agarose as an anticonvection medium is its ability to exhibit high gel strength at low concentrations (< 6%). Gel strength is defined as the force, expressed in g/cm^2, that must be applied to fracture an agarose gel of a standard concentration. As there are several test methods used to measure gel strength, a direct comparison of gel strength values between different manufacturers is sometimes difficult. The gel strength of a specific lot of agarose decreases over time because of spontaneous hydrolysis of the agarose polysaccharide chains. This loss of gel strength can be particularly noticeable after 5 years from the manufacturing date.

2 Preparation of agarose gels

2.1 Buffer systems

The most commonly used technique for DNA separation is submerged horizontal electrophoresis in 0.5–6.0% agarose, usually in TAE or TBE buffer (*Table 1*). By varying the agarose concentration and/or buffer, it is possible to separate reliably double-stranded DNA in the size range of approximately 20–50 000 bp. For analytical purposes, most low EEO agaroses are useful for DNA separations. When recovering DNA, it is best to use an agarose specially certified to yield biologically active DNA.

Table 1 Buffer formulations

Buffer	Recipe	Concentration of the 1× buffer
50× TAE	242 g Tris base 57.1 ml glacial acetic acid 18.61 g Na$_2$EDTA.2H$_2$O Add distilled water to 1 l	40 mm Tris base, 40 mm acetic acid, 1 mm EDTA
5× TBE	54.0 g Tris base 27.5 g boric acid 3.72 g Na$_2$EDTA.2H$_2$O Add distilled water to 1 l	89 mm Tris base, 89 mm boric acid, 2 mm EDTA
10×TPE	108 g Tris base 15.5 ml 85% phosphoric acid 7.44 g Na$_2$EDTA.2H$_2$O Add distilled water to 1 l	90 mm phosphoric acid, 2 mm EDTA

2.1.1 The choice between TAE and TBE

If the DNA is less than 12–15 kb and will not be recovered from the gel, then either TAE or TBE buffer at a standard concentration can be used. For larger DNA, the best choice is TAE in combination with a low field strength (1–2 V/cm). During these extended electrophoretic runs, the larger gel porosity, lower EEO and low field strength reduce the tendency of the DNA to smear.

TBE buffer is preferred for separation of small DNA molecules (< 1 kb) when recovery is not required because the interaction between TBE and agarose results in a smaller apparent pore size. This tighter gel reduces the broadening of DNA bands owing to dispersion and diffusion.

Whichever buffer is used, the depth over the gel in a horizontal electrophoretic system should be 3–5 mm—with less buffer the gel may dry out during electrophoresis. Excessive buffer depth decreases DNA mobility, promotes band distortion and can cause excessive heating within the system.

2.1.2 Buffer depletion

The rate of buffer depletion is influenced by the buffer used and its buffering capacity. A 0.5× TBE buffer has greater buffering capacity than a 1× TAE buffer at the pH used because the pK_a of borate is closer to the initial buffer pH than is the pK_a of acetate. Standard-sized electrophoresis chambers (15 × 30 cm) with a 1.5-2 l capacity will tolerate 40–50 W·h before buffer depletion, and buffer depletion will not occur in mini-electrophoresis chambers for 10–13 W·h. Consult the manufacturer's instructions provided with the electrophoresis chamber for specific values.

Evidence of buffer depletion is melting gel, smearing of DNA and/or overheating. These effects of buffer depletion and development of a pH gradient can be reduced by recirculating the buffer. This is usually necessary only when electrophoresis is performed for extended times or the electrophoresis buffer has a low buffering capacity.

2.1.3 Other buffering systems

Tris-phosphate buffer (TPE: see *Table 1*) may also be used for DNA electrophoresis. Like TBE buffer, TPE has a high buffering capacity but it does not interfere with DNA recovery procedures. However, TPE is unsuitable if the recovered DNA is be used in an enzymatic reaction that is sensitive to contaminating phosphate ions.

2.2 Dissolving agarose

Agarose undergoes a series of processes when it is dissolved: dispersion, hydration and melting/dissolution. Dispersion simply refers to the separation of the particles by the buffer without clumping. Clumping occurs when the agarose starts to dissolve before it is completely dispersed, coating itself with a gelatinous layer which inhibits the penetration of water and keeps the powder inside from dispersing. Dissolution then becomes a long process. Hydration is the surrounding of agarose particles by a solution (e.g. water or running buffer). Problems are sometimes encountered with hydration when using a microwave oven to dissolve agarose. In part, this occurs because hydration is time dependent and microwave ovens bring the temperature up rapidly. The problem is exacerbated by the fact that the agarose is not being agitated to help dilute the highly concentrated solution around each particle, and dissolution is slowed.

The final stage in dissolving the agarose is the melting and dissolution of a high concentration gel. Melting can be done in either a microwave oven or on a hot plate. As the particles hydrate, they become small, highly concentrated gels. Since the melting temperature of a standard agarose gel is about 93 °C, merely heating a mixture to 90 °C will not completely dissolve agarose. Even low-melting-temperature agaroses should be boiled when preparing gels to ensure that the solution is degassed and all agarose is fully dissolved.

Protocol 1

Dissolving agarose

Equipment and reagents

Caution: **Always wear eye protection, and guard yourself and others against scalding solutions.**

- Microwave oven or hotplate
- Agarose powder

- $1\times$ TAE, $1\times$ TBE or $0.5\times$ TBE electrophoresis buffer (see *Table 1*)

A. For a gel concentration of $\leq 2\%$ (w/v)

1   Choose a beaker that is two to four times the volume of the solution.

2   Add the appropriate amount of unchilled or chilled[a] $1\times$ or $0.5\times$ buffer and a stir bar to the beaker.

3   Sprinkle in the premeasured agarose powder while the buffer is being rapidly stirred.

4   Remove the stir bar. Weigh the beaker and solution and cover with plastic wrap. Pierce a small hole in the plastic wrap for ventilation.

5   Heat the beaker in the microwave oven on high power until bubbles appear. **Caution: The microwaved solution may become superheated and foam over when agitated.**

6   Remove the beaker from the microwave oven. Swirl gently to resuspend any settled powder and gel pieces.

7   Reheat the beaker on high power until the solution comes to a boil. Hold at the boiling point for 1 min or until all of the particles are dissolved, then remove the beaker from the microwave oven. Swirl gently to thoroughly mix the agarose solution.

8   After dissolution, add sufficient hot distilled water to return to the initial weight.

9   Mix thoroughly and cool the solution to 50–60°C prior to casting.

B. For a gel concentration of $> 2\%$ (w/v)

1   Follow steps 1–4 of *Protocol 1A* then leave to agarose to soak in the buffer for 15 min.

2   If the gel concentration is $> 4\%$ then heat the beaker in the microwave oven on medium power for 1 min, remove from the oven and leave on the bench for 15 min. If the concentration is 2–4% then move directly to step 3. **Caution: The microwaved solution may become superheated and foam over when agitated.**

3   Heat the beaker in the microwave oven on medium power for 2 min.

4   Remove the beaker from the microwave oven. Swirl gently to resuspend any settled powder and gel pieces.

5   Follow steps 7–9 of *Protocol 1A* to complete preparation of the gel.

[a] Prechilled buffer gives better dispersion of some agaroses designed especially for high gel concentrations: check the manufacturer's instructions.

2.3 Casting an agarose gel

For optimal resolution, cast horizontal gels 3–4mm thick at the appropriate concentration (see *Table 2*). The appropriate volume of gel solution can be estimated by measuring the surface area of the casting chamber, then multiplying by the gel thickness. Thinner gels can be cast on a plastic film (e.g. BioWhittaker Molecular Applications GelBond film) and/or in a vertical electrophoresis apparatus.

Table 2 Suggested agarose concentrations

		Final agarose concentration (% w/v)	
Agarose	**Resolution (bp)**	**1× TAE buffer**	**1× TBE buffer**
NuSieve 3:1[a]	500–1000	3.0	2.0
	100–500	4.0	3.0
	10–100	6.0	5.0
NuSieve GTG[a]	500–1000	2.5	2.0
	150–700	3.0	2.5
	100–450	3.5	3.0
	70–300	4.0	3.5
	10–100	4.5	4.0
	8–50	5.0	4.5
MetaPhor[a]	150–800	2.0	1.8
	100–600	3.0	2.0
	50–250	4.0	3.0
	20–130	5.0	4.0
	<80	Not applicable	5.0
SeaPlaque and	500–25 000	0.75	0.70
SeaPlaque GTG	300–20 000	1.00	0.85
	200–12 000	1.25	1.00
	150–6000	1.50	1.25
	100–3000	1.75	1.50
	50–2000	2.00	1.75
SeaKem GTG and	1000–23 000	0.60	0.50
SeaKem LE	80–10 000	0.80	0.70
	400–8000	1.00	0.85
	300–7000	1.20	1.00
	200–4000	1.50	1.25
	100–3000	2.00	1.75

[a] Preparing NuSieve 3:1, NuSieve GTG and MetaPhor agarose gels at less than 2% is not recommended because these gels are more difficult to handle.

Protocol 2

Casting a horizontal agarose gel

Equipment and reagents

- Horizontal electrophoresis apparatus plus accessories (e.g. power supply, gel plate, comb[a])
- Agarose solution (prepared as described in *Protocol 1*)
- Electrophoresis buffer (see *Table 1*)

Method

1 Allow the agarose solution to cool to 50–60 °C.

2 While the solution is cooling, assemble the gel casting tray. Level the tray and check the teeth of the comb for residual dried agarose. Dried agarose can be removed by scrubbing the comb with a lint-free tissue soaked in hot distilled water. Allow a small space (< 0.5–1.0 mm) between the bottom of the comb teeth and the casting tray.

3 Pour the agarose solution into the gel tray.

4 Position the comb and allow the agarose to gel at room temperature for 30 min.[b]

5 Once the gel has set, flood with running buffer and slowly remove the comb.

6 Place the gel casting tray into the electrophoresis chamber and fill the chamber with electrophoresis buffer until the buffer reaches 3–5 mm over the surface of the gel.

7 Gently flush out the wells with electrophoresis buffer using a Pasteur pipette to remove loose gel fragments prior to loading the samples.

8 Load the DNA and electrophorese (see Section 2.4).

[a] The thickness of the comb in the direction of the electric field can affect the resolution of the gel. A thin comb (1 mm) results in sharper DNA bands. With a thicker comb, a greater volume can be added to the well but the separated DNA bands may be broader.

[b] Low-melting-temperature and intermediate-melting-temperature agaroses require an additional 30 min of gelling at 4 °C to obtain the best gel handling.

2.4 Loading and running DNA in an agarose gel

Variable amounts of DNA can be loaded per well, the critical factor being the quantities of DNA in the bands of interest after the gel has been run. The following points should be considered:

• Well volume

• Fragment size—the capacity of the gel drops sharply as the fragment size increases, even over a few kb

• Distribution of fragment sizes

• Voltage gradient—higher voltage gradients are better suited for DNA fragments < 1 kb, whereas lower voltages are better suited for fragments > 1 kb

The smallest amount of DNA that can be reliably detected in a single band is approximately 10 ng with ethidium bromide staining and as low as 60 pg with a special stain such as GelStar nucleic acid stain (BioWhittaker Molecular Applications) or SYBR Green I stain (BioWhittaker Molecular Applications). The maximum amount of DNA that can be loaded without losing band definition is about 100 ng. Overloaded DNA results in trailing and smearing, a problem that becomes more severe as the size of the DNA increases.

2.4.1 Loading buffers

Gel loading buffers serve three purposes in DNA electrophoresis:

- They increase the density of the sample, ensuring that the DNA settles evenly into the well
- They add colour to the sample and so simplify loading
- They contain mobility dyes which migrate in an electric field towards the anode at predictable rates: this enables the electrophoretic process to be monitored

A number of loading buffers are commonly used for agarose gel electrophoresis (*Table 3*). They are usually prepared as 6× or 10× stock solutions. Alkaline loading buffer is used when performing alkaline gel electrophoresis.

To increase the sharpness of DNA bands, use Ficoll 400 as a sinking agent instead of glycerol. The inclusion of the lower molecular weight glycerol in the loading buffer does not always prevent the DNA from streaming up the sides of the well before electrophoresis has begun and can result in a U-shaped band (7). In TBE gels, glycerol also interacts with borate which can alter the local pH. A loading buffer with too high an ionic strength causes bands to be fuzzy. Ideally, the DNA sample should be resuspended in the same solution as the running buffer. If this is not possible, use a sample buffer with a lower ionic strength than the running buffer.

The zone of highest resolution for DNA separation in an agarose gel is in the middle third of the gel. Migration data for the common mobility dyes bromophenol blue and xylene cyanol are shown in *Table 4*. These data relate the mobility of the dyes to the migration rate of double-stranded DNA and are of assistance in determining optimal run times.

The voltage gradient is determined from the distance between the electrodes, not the gel length. If the voltage is too high, band streaking, especially for DNA ≥ 12–15 kb, may result. When the voltage is too low, the mobility of small

Table 3 Loading buffer recipes

Loading buffer	Recipe for 6× buffer	Storage temperature
Sucrose based	40% (w/v) sucrose 0.25% (w/v) bromophenol blue 0.25% (w/v) xylene cyanol	4°C
Glycerol based	30% (v/v) glycerol 0.25% (w/v) bromophenol blue 0.25% (w/v) xylene cyanol	4°C
Ficoll based	15% (w/v) Ficoll (type 400) 0.25% (w/v) bromophenol blue 0.25% (w/v) xylene cyanol	Room temperature
Alkaline buffer	300 mM NaOH 6 mM EDTA 18% (w/v) Ficoll (type 400) 0.15% (w/v) bromocresol green 0.25% (w/v) xylene cyanol	4°C

Table 4. Dye mobilities for DNA gels

| Agarose (% w/v) | Running position of the dye markers (bp)[a] | | | |
| | 1× TAE buffer | | 1×TBE buffer | |
	XC[b]	BPB[b]	XC[b]	BPB[b]
NuSieve 3:1 agarose				
2.5	950	130	700	70
3.0	650	80	500	40
4.0	350	40	250	20
5.0	200	30	140	8
6.0	120	20	90	4
NuSieve GTG agarose				
2.5	750	175	460	75
3.0	400	120	210	35
4.0	115	<20	150	<20
5.0	100	<20	80	<20
6.0	85	<20	50	<20
MetaPhor agarose				
2.0	480	70	310	40
3.0	200	40	140	35
4.0	120	35	85	30
5.0	85	30	60	15
SeaPlaque and SeaPlaque GTG agaroses				
0.75	4000	500	2850	280
1.00	2300	350	1700	180
1.50	1000	150	700	70
2.00	550	60	400	30
2.50	320	30	250	10
SeaKem GTG and SeaKem LE agaroses				
0.30	24 800	2900	19 400	2850
0.50	11 000	1650	12 000	1350
1.00	6100	500	4100	400
1.50	2800	300	1800	200
2.00	1300	150	850	70
SeaKem Gold agarose				
0.30	24 800	3550	19 000	2550
0.50	12 200	2050	9200	1500
1.00	6100	760	4000	500
1.50	2600	400	1900	250
2.00	1500	250	1000	100

[a] The data relate the mobilities of the dyes to the migration rate of double-stranded DNA.

[b] Abbreviations: BPB, bromophenol blue; XC, xylene cyanol.

(≤ 1 kb) DNA is reduced and band broadening occurs owing to dispersion and diffusion. *Table 5* provides a quick reference for the optimal voltage and buffer for electrophoresis of DNA molecules of various sizes.

2.5 Detection of DNA in agarose gels

2.5.1 Detecting DNA with ethidium bromide

Ethidium bromide is a fluorescent dye which detects both single- and double-stranded DNA. However, the affinity for single-stranded DNA is relatively low

Table 5 Recommended electrophoresis conditions for different applications

| DNA size (kb) | Recommended voltage (V/cm) | Recommended buffer | |
		Analytical gel[a]	Preparative gel[a]
≤ 1.0	5	TBE	TAE
1.0–12.0	4–10	TAE or TBE	TAE
> 12.0	1–2	TAE	TAE

[a] An analytical gel is used simply to determine the sizes of the DNA molecules in the samples; a preparative gel is used both to determine the sizes of the molecules and to purify one or more molecules using the recovery techniques described in Chapter 6.

Table 6 Sensitivities of stains used to visualize DNA in agarose gels

| Stain | Detection limits (ng DNA) | |
	Double-stranded DNA	Single-stranded DNA
Ethidium bromide	1–5	5
Silver stain	1	5
Methylene blue	40–200	Not applicable
Acridine orange[a]	50	100

[a] Acridine orange distinguishes single-stranded from double-stranded DNA: when *Protocol 6* is used, double-stranded DNA fluoresces green under shortwave UV irradiation and single-stranded DNA fluoresces red.

compared with that for double-stranded DNA (*Table 6*). Ethidium bromide contains a planar group which intercalates between the bases of DNA, resulting in an increase in fluorescence yield when the DNA/ethidium bromide complex is exposed to UV radiation. At 254 nm, UV is absorbed by the DNA and transmitted to the dye; at 302 nm and 366 nm, UV is absorbed by the bound dye itself. In both cases, the energy is re-emitted at 590 nm in the red–orange region of the visible spectrum.

Protocol 3

Staining DNA in agarose gels with ethidium bromide

Equipment and reagents

Caution: Ethidium bromide is a powerful mutagen. Gloves should be worn when handling solutions of this dye and stained gels. Appropriate eye and skin protection should be worn when observing gels on a UV transilluminator. See Chapter 1, Sections 3.3 and 3.4.

- Staining vessels larger than the gel
- UV transilluminator
- Electrophoresis buffer (see *Table 1*) or distilled water

- Ethidium bromide stock solution (10 mg/ml). Add 1 g ethidium bromide to 100 ml distilled water and stir on a magnetic stirrer for several hours. Transfer the solution to a dark bottle and store at room temperature

A. Procedure for post-staining gels

1 Prepare enough working solution of ethidium bromide (0.5–1.0 mg/ml of ethidium bromide in distilled water or electrophoresis buffer) to cover the surface of the agarose gel.

2 Remove the gel from the electrophoresis chamber.

3 Submerge the gel for 20 min[a] in the ethidium bromide solution.

4 Remove the gel and destain by submerging for 20 min in a new container filled with distilled water.

5 Repeat step 4, using fresh distilled water.[b]

6 Gels can be viewed with a hand-held or tabletop UV emitter.

7 Decontaminate the used ethidium bromide solutions as described in *Appendix 2, Protocol 5*.

B. Procedure for inclusion of ethidium bromide in the agarose gel

1 Prepare the agarose solution as described in *Protocol 1*.

2 While the agarose solution is cooling, add ethidium bromide to a final concentration of 0.1 to 0.5 μg/ml. Slowly swirl the solution and then cast the gel as described in *Protocol 2*.

3 Add ethidium bromide to the electrophoresis buffer to a final concentration of 0.5 μg/ml. Load the DNA samples and run the gel.

4 Destain the gel by submerging in distilled water for 20 mins. Repeat with fresh distilled water.[b]

5 Gels can be viewed with a hand-held or tabletop UV emitter during or after electrophoresis.

6 Decontaminate the used ethidium bromide solutions as described in *Appendix 2, Protocol 5*.

[a] For gel concentrations of 4% or greater, this time may need to be doubled.

[b] The destaining steps can be repeated more than twice if the background is still too high.

For optimal resolution, sharpest bands and lowest background, stain the gel with ethidium bromide following electrophoresis. Ethidium bromide can also be included in the gel and electrophoresis buffers (0.5 mg/ml) with only a minor loss of resolution. The electrophoretic mobility of DNA will be reduced by approximately 15%.

2.5.2 Detecting DNA by silver staining

DNA in agarose gels can be stained with silver in a manner similar to staining DNA in polyacrylamide gels (8). However, because agarose has a slight negative charge, commercial reagents that work only on polyacrylamide gels will not work with agarose gels. A kit for agarose gels is available (Bio-Rad Silver Stain

Plus), or *Protocol 4* can be followed. Even when the procedure is optimized for DNA in agarose gels, silver staining of DNA will only give a sensitivity similar to ethidium bromide. Nevertheless, silver staining is the only available method when a permanent record of the gel is required.

If it is necessary to detect low levels of DNA (< 1 ng), the use of the highly sensitive nucleic acid detection dyes, SYBR Green I or GelStar stain (Bio-Whittaker Molecular Applications), is preferable. New varieties of these sensitive dyes are regularly introduced, and should be used in accordance with the manufacturer's instructions.

Protocol 4

Silver staining DNA in agarose gels

Equipment and reagents

Caution: Wear the necessary protective equipment and follow your local regulations for disposal of silver solutions.

- GelBond film (BioWhittaker Molecular Applications)
- Staining vessels larger than the gel
- Fixative solution: mix 500 ml methanol, 120 ml glacial acetic acid and 50 g glycerol and then bring the volume to 1 l with distilled water; prepare fresh
- Solution A: dissolve 50 g sodium carbonate in 800 ml distilled water then bring the volume to 1 l with more distilled water; this is stable for 2–3 weeks at room temperature

- Solution B: dissolve 2 g ammonium nitrate + 2 g silver nitrate + 10 g tungstosilic acid (Sigma T2786) in 800 ml distilled water; add 8 ml 37% (v/v) formaldehyde and bring the volume to 1 l with distilled water; this is stable for 1 week when stored in the dark at room temperature
- Stop solution: 1% (v/v) glacial acetic acid in distilled water; prepare fresh

Method

1 Cut the GelBond film to the inner dimension of your casting tray.[a]

2 Spread a few drops of distilled water on the surface of the casting tray. Lay the cut sheet of GelBond film onto the casting tray with the hydrophilic side up.

3 Cover the film with a sheet of blotting paper and firmly roll with a rubber roller or wipe with tissues to squeeze out any air bubbles and excess fluid from beneath the film. Wipe off any excess liquid at the edges.

4 Prepare a gel 3 mm thick, as described in *Protocols 1* and *2* (a thin gel is necessary to ensure a low background).

5 Electrophorese the DNA as described in *Protocol 2*.

6 Dry the gel.

7 Fix the gels for 30 min in the fixative solution,[b] using sufficient solution to cover the gel to a depth of 5 mm.

Protocol 4 continued

8   Wash the gel for 3×20 min in distilled water.

9   Vigorously stir solution A on a magnetic plate, then mix equal volumes of solutions A and B to give the staining solution.

10 Place the gel in a glass vessel and add enough staining solution to submerge to a depth of 5 mm. Leave the gel in the stain until the bands appear.

11 Transfer the gel to the stop solution for 5 min.

12 Rinse the gel in distilled water and allow to air dry.

[a] Gelbond film is recommended for easy handling of the gel and to prevent curling during drying.

[b] If using Bio-Rad's Silver Stain Plus Kit, follow the manufacturer's instructions instead of steps 7–12.

2.5.3 Detecting DNA by staining with methylene blue or acridine orange

Both methylene blue and acridine orange can be used to stain DNA in agarose gels. Methylene blue (*Protocol 5*) is non-carcinogenic and does not require the use of UV for detection. This can be an advantage when DNA damage needs to be avoided (9). Methylene blue is approximately 40 times less sensitive than ethidium bromide and should be used only when there are large amounts of DNA (> 40 ng per band). If it is necessary to differentiate between single-stranded and double-stranded DNA in a gel, acridine orange can be used (10). When *Protocol 6* is followed, double-stranded DNA fluoresces green under short-wave UV irradiation and single-stranded DNA fluoresces red. When increased ratios of dye to DNA are used, all molecules fluoresce green. This stain is not ideal for everyday use because the background fluorescence can often be high, leading to low signal-to-noise ratios and decreased sensitivity of detection.

Protocol 5

Detecting DNA with methylene blue

Equipment and reagents

- Methylene blue stain solution: 0.02% (w/v) methylene blue in distilled water
- Staining vessels larger than the gel

Method

1   Remove the gel from the electrophoresis apparatus and place in sufficient staining solution to cover to a depth of 5 mm.

2   Stain for 15 min then destain gel for 15 min in distilled water.

Protocol 6

Detecting DNA with acridine orange

Equipment and reagents

- Enamel staining container larger than the gel
- UV transilluminator
- Acridine orange stock solution: 10 mg/ml acridine orange in distilled water; store protected from light at 4°C

Method

1 Dilute the acridine orange stock solution to 30 μg/ml in distilled water.

2 Remove the gel from the electrophoresis apparatus and place in the enamel container. Cover to a depth of 5 mm with the diluted acridine orange solution.

3 Stain the gel in the dark for 30 min then destain, again in the dark, for 30 min in distilled water. Removal of high background fluorescence can be achieved by destaining overnight at 4°C in the dark.[a]

4 View under 300 nm UV radiation.[b]

5 Clean the enamel staining container by washing with 95% (v/v) ethanol.

[a] If destaining is incomplete, the background will appear very green and the red bands will appear dull or even black. Faint bands may not be visible at this stage.

[b] Photography of gels stained with acridine orange is best carried out with a red photographic filter (Wratten No. 24). The red filter enhances the discrimination of the red bands against the green background.

2.6 Alkaline gel electrophoresis

In most cases, DNA is electrophoresed in an agarose gel as double-stranded molecules, in other words under non-denaturing conditions. However, for some applications, sodium hydroxide is added to an agarose gel. The resulting alkaline gel, when combined with an alkaline gel buffer, promotes and maintains denaturation of DNA into single-stranded molecules during electrophoresis. The buffering capacity and resistance of the gel buffer are very low and these factors require that a low voltage is used for the electrophoresis. The relevant applications are:

- To analyse single-stranded DNA
- To size cDNA strands in DNA–RNA hybrids
- To size first and second strands of cDNA (Volume II, Chapter 3)
- To check for nicking activity of enzyme preparations used in cloning and other procedures

Protocol 7

Alkaline gel electrophoresis

Equipment and reagents

- Standard-melting-temperature agarose powder
- Alkaline loading buffer (6× stock solution contains 300 mM NaOH, 6 mM EDTA, 18% [w/v] Ficoll 400, 0.15% [w/v] bromocresol green,[a] 0.25% [w/v] xylene cyanol)

- Horizontal electrophoresis apparatus plus accessories (e.g. power supply, gel plate, comb)
- Alkaline electrophoresis buffer (50 mM NaOH, 1 mM EDTA), prepared fresh

Method

1 Prepare the agarose solution in distilled water (see *Protocol 1*).[b]

2 Cool the solution to 50–60 °C. Add NaOH to a final concentration of 50 mM and EDTA, pH 8.0, to a final concentration of 1 mM.

3 Cast the gel as described in *Protocol 2* and allow to set at room temperature.

4 Place the gel in the electrophoresis apparatus and add alkaline electrophoresis buffer so the gel is covered to a depth 3–5 mm.

5 Add 0.2 volumes of 6× alkaline loading buffer to DNA samples and load these into the wells of the gel.

6 Electrophorese at 0.25 V/cm until the bromocresol green has migrated out of the well and into the gel. Alkaline gels draw more current than neutral gels at comparable voltages and will heat up during electrophoresis.

7 Place a glass plate on top of the gel. This will keep the bromocresol green from diffusing into the buffer.

8 Electrophorese at 0.25–1.80 V/cm until the dye has migrated approximately two-thirds of the length of the gel.

9 Stain the gel in a 0.5 μg/ml ethidium bromide solution (*Protocol 3*) or with SYBR Green I (see Section 2.5.2).

[a] Bromocresol green is used instead of bromophenol blue because it displays a more vivid colour under alkaline conditions.

[b] Do not add NaOH to a hot agarose solution, as it causes hydrolysis of the polysaccharide.

2.7 Drying an agarose gel

Occasionally, agarose gels are dried as a preliminary to autoradiography or in-gel hybridization, or as a preservation measure. Gels can be dried with or without a vacuum dryer, and can be attached, or not attached to a support material. When using a vacuum drying system, the process is essentially the same as for a polyacrylamide gel except that high heat cannot be used.

2.7.1 Drying an agarose gel without a vacuum gel dryer

To dry an agarose gel at room temperature (*Protocol 8*) or in a hot-air oven (*Protocol 9*), the gel must have been cast on to a support material such as GelBond film (BioWhittaker Molecular Applications). This prevents the gel from shrinking during the drying process.

Protocol 8

Drying an agarose gel at room temperature

Reagents

• Agarose gel cast on GelBond film

Method

1 Remove the gel from the electrophoresis apparatus and cut off 0.5–1 cm of agarose from each side of the gel, leaving the GelBond film exposed.

2 Clamp the exposed GelBond film, gel side up, to a glass plate or particle board using clamps or elastic bands. This will prevent the gel from curling during drying.

3 Allow to dry on the bench top for at least 8 h.

Protocol 9

Drying an agarose gel in a forced hot-air oven

Equipment and reagents

• Agarose gel cast on GelBond film

• Forced hot-air oven

Method

1 Set the temperature of the forced hot-air oven to 55 °C.

2 Prepare the gel as described in steps 1 and 2 of *Protocol 8*.

3 Place in the oven and allow to dry for 1–4 h, depending on the thickness of the gel.

2.7.2 Drying an agarose gel with a vacuum gel dryer

Standard-melting-temperature agarose gels can be dried in a standard vacuum gel dryer. This method has the advantage that the gels do not necessarily need to be cast on a support material, although this can be done during the drying process (11) and provides a stable support for the brittle, dried gel. Caution should be exercised when using this method to dry gels that contain radioactivity because some of the radioactivity is transferred from the gel to the vacuum trap.

Protocol 10

Drying an agarose gel with a vacuum dryer

Equipment and reagents

- Vacuum gel dryer with variable temperature control
- Whatman 3 MM chromatography paper
- GelBond film (BioWhittaker Molecular Applications) or other support material (optional)

Method

1 Set the temperature on the vacuum gel dryer to 60 °C. A higher temperature may cause the gel to melt.

2 Place three sheets of chromatography paper onto the vacuum gel dryer.

3 Remove the gel from the electrophoresis apparatus and place onto the chromatography paper. If the gel is adhered to a support film, then the gel must be placed between the support and the vacuum source.

4 Cover the gel with plastic wrap, turn on the vacuum and dry for 1–2 h.

5 After drying, remove the chromatography paper from the agarose gel by moistening the paper with distilled water and carefully peeling it off. Note that agarose gels do not adhere to chromatography paper as well as polyacrylamide gels and may spontaneously become unstuck.

Note

Agarose gels not cast on GelBond film prior to electrophoresis can be permanently adhered to GelBond film during the drying process. After removing the gel from the electrophoresis apparatus, place it with wells facing downwards onto the three sheets of chromatography paper. Place a sheet of GelBond film onto the gel, hydrophilic side in contact with the gel, and then cover the gel and GelBond film with plastic wrap. Turn on the vacuum without heat for 30 min, then set the dryer temperature to 60 °C and dry for 40 min. After drying, treat the gel as described in step 5.

3 Vertical agarose gels and fast-running horizontal gels

3.1 Vertical gels

Agarose gels can be cast in standard vertical gel electrophoresis systems. The advantages of vertical agarose gels are faster run times and increased resolution; both are a result of the higher voltages that can be used. Agarose gels in a vertical format require careful preparation because the agarose can gel during the casting process, and agarose is not as elastic as polyacrylamide, which makes removal of the comb a more delicate process. A 2-bp resolution is possible for double-stranded DNA, but the upper limit of separation is lower than in standard horizontal agarose gels. Follow the steps in *Protocol 11* to cast a vertical agarose gel.

Protocol 11

Casting a vertical agarose gel

Equipment and reagents

- Vertical electrophoresis apparatus plus accessories (e.g. power supply, gel plates, combs and side-spacers)
- Whatman 3 MM chromatography paper
- Clamps
- Silicone tubing or electrical tape

- Two 60 ml syringes with 16-gauge needles
- Heat gun or 55 °C oven
- Agarose solution (see *Protocol 1*)
- Electrophoresis buffer (see *Table 1*)

A. Cassette assembly

1   Clean the glass plates with soap and water, rinse with distilled water and dry. Wipe the plates with ethanol and a lint-free tissue.

2   Unlike polyacrylamide gels, agarose gels do not adhere to glass plates and may slide out during electrophoresis. To prevent this, frosted glass plates can be used or this procedure followed:

  (a) Place two side-spacers on the back plate.

  (b) For 1-mm gels, cut a strip of Whatman 3MM chromatography paper (1 mm thick and 5–10 mm wide) long enough to fit between the two spacers and wet with electrophoresis buffer. Place these strips at the bottom of the back plate in contact with the spacers on either side.

3   Clamp the front and back glass plates together. Use the manufacturer's casting apparatus or seal the cassette against leaks with silicone tubing or tape.

B. Follow one of the methods below to seal the cassette prior to casting the gel

Silicone tubing method

1   Use silicone tubing which is the same diameter as the spacer thickness. Cut a piece long enough to extend along the bottom and up both sides of the cassette.

2   Place the tubing across the bottom of the back plate below the blotting paper strip.

3   Place the top plate over the bottom plate and clamp the plates together at the bottom.

4   Run the tubing up either side of the plates and finish clamping the plates together.

Tape method

1   Place the top plate over the bottom plate and tape the sides with separate pieces of tape.

2   Tape the bottom of the cassette with a separate piece of tape. This way the tape on the bottom can be removed for electrophoresis without disturbing the tape at the sides of the gel.

3   Clamp the plates together.

C. Casting a vertical agarose gel

1   Prepare the agarose solution in 1× TAE or TBE buffer (see *Protocol 1*)

2   Pre-warm the assembled cassette and a 60-ml syringe for 15 min by placing these in a 55 °C oven or by using a heat gun.

3   Cool the dissolved agarose to 60 °C and pour into the pre-warmed 60 ml syringe fitted with a 16-gauge needle.

4   Wedge the needle tip between the plates in the upper corner of the cassette with the needle opening directed toward the back plate and inject the agarose solution at a moderate, steady rate. Keep a constant flow to prevent air bubble formation. Angle the cassette while pouring so the agarose solution flows down one side spacer, across the bottom and up the other side. Fill until the agarose solution is just above the glass plates.

5   Insert one end of the comb, then slowly insert the rest of the comb until the teeth are at an even depth. Insert the comb into the agarose to the minimal depth necessary to accommodate your samples.

6   Place extra clamps on the side of the glass plates to hold the comb in place.

7   Cool the gel at room temperature for 15 min, then place the gel at 4 °C for 20 min.

8   Remove the clamps at the top of the gel, and remove any excess agarose with a scalpel or razor blade.

9   Squirt electrophoresis buffer into the spaces between the comb and the gel, then slowly and gently lift the comb straight up. Allow air or buffer to enter the well area to release the vacuum that forms between the agarose and the comb. The wells can be further cleaned by flushing with running buffer.

10   The gel can be stored overnight in a humidity chamber or in a sealed bag with a buffer-dampened paper towel.

D. Preparing the gel for electrophoresis

1   Remove the silicone tubing or tape at the bottom of the cassette.[a] If you have placed chromatography paper between the plates at the bottom, it is not necessary to remove it as it will not interfere with electrophoresis.

2   Place the cassette in the electrophoresis chamber at an angle to minimize the number of bubbles that can collect in the well area.

3   Rinse out the well area with a syringe.

4   Load the DNA samples and perform the electrophoresis using the conditions described in *Table 7*.

[a] Because agarose does not adhere well to glass, leave as many clips in place as possible. For some electrophoresis chambers, it is helpful to seal the spacers at the top of the gel with molten agarose.

Table 7 Vertical electrophoresis conditions

Parameter	Condition/setting
Run time	1.0–1.5 h
Gel concentration	3–4%
Gel length	16–26 cm
Gel thickness	1 mm
Gel buffer	1× TAE or 1× TBE
Electrophoresis buffer	1× TAE or 1× TBE
Voltage[a]	17 V/cm
Temperature	Ambient

[a] V/cm is determined by dividing the total voltage by the distance between the electrodes.

3.2 Fast-running horizontal gels

A standard horizontal system can be used to achieve a resolution nearly as good as with a vertical agarose gel but without the difficulties involved with casting. The only special apparatus required is a chilling, recirculating water bath. Either a commercially available system can be used, or a home-made version made consisting of an ice-water bath, aluminium tubing, Tygon tubing and a peristaltic pump.

Protocol 12

Fast-running horizontal gels

Equipment and reagents

- Horizontal electrophoresis chamber to accommodate a 20 cm gel, plus accessories (e.g. power supply, gel plate, comb)
- Recirculating–chilling water bath
- TAE or TBE buffer (see *Table 1*)

Method

1 Prepare a 3–4% MetaPhor agarose gel (BioWhittaker Molecular Applications) in 1× electrophoresis buffer.

2 Cast a 3-mm thick, 20-cm-long agarose gel.

3 Allow the gel to solidify at room temperature, then leave at 4 °C for 30 min before placing in the electrophoresis chamber.

4 Add 0.5× TBE or 1× TAE to a depth of 3 mm over the surface of the gel.

5 Load 20–50 ng of DNA per well. If the DNA concentration cannot be estimated, load varying amounts of the sample. The sharpest bands are seen with small (5–10 μl) loading volumes.

6 Run the gel at 17 V/cm.

7 When the sample has left the well and moved into the gel, begin recirculating and chilling the electrophoresis buffer with a recirculator–chiller water bath.[a]

Protocol 12 continued

8 Run the gel for < 1.5 h using the conditions described in *Table 8*.

[a] This protocol cannot be used with just a peristaltic pump in the cold room, because the gel will melt.

Table 8 Fast running horizontal electrophoresis conditions

Parameter	Condition/setting
Run time	1.5 h
Gel concentration	3–4%
Gel length	20 cm
Gel thickness	3 mm
Gel buffer	1× TAE or 1× TBE
Electrophoresis buffer	0.5× TAE or 0.5× TBE
Voltage[a]	17 V/cm
Temperature	Ambient

[a] V/cm is determined by dividing the total voltage by the distance between the electrodes.

4 Electrophoresis of RNA

RNA requires different techniques from DNA for electrophoretic separations. Because single-stranded RNA can form secondary structures, it must be electrophoresed in a denaturing system. RNA must be denatured prior to loading the gel and the electrophoresis conditions should maintain the denatured state.

4.1 Denaturing systems

When choosing a denaturing system, the purpose of the experiment and the size of the RNA being separated should be kept in mind. The most frequently used denaturants for agarose gel electrophoresis of RNA are formaldehyde (12–14) and glyoxal plus DMSO (10). Protocols are given in ref. 15 and are described below. In each system, fully denatured RNA migrates through the agarose gel in a linear relation to the logarithm of its molecular weight. Methylmercuric hydroxide (CH_3HgOH) is the most efficient RNA denaturant; it reacts with the imino bonds of uridine and guanosine and disrupts all secondary structure (16). Unfortunately, methylmercuric hydroxide is extremely hazardous to use (17), so despite its high efficiency in denaturing RNA it is by far the least popular denaturation method and we do not recommend it.

The resolving powers of the glyoxal/DMSO and formaldehyde buffer systems are nearly equal for most purposes. However, for detection by Northern analysis, glyoxal/DMSO denaturant is preferable because these gels tend to produce somewhat sharper bands than the formaldehyde system. Glyoxal gels require more care to run than formaldehyde gels and, because of the lower buffering capacity of glyoxal, these gels must be run at lower voltages than formaldehyde gels; the buffer must be recirculated to avoid creating hydrogen ion gradients during electrophoresis (5). If the pH of the buffer rises above 8.0, glyoxal dissociates from RNA, causing the RNA to renature and migrate in a unpredictable manner (10). For staining purposes, either denaturant can be used. Ethidium bromide, GelStar nucleic acid stain (BioWhittaker Molecular Applications) and SYBR Green II gel stain (BioWhittaker Molecular Applications) all bind better to formaldehyde-denatured RNA than glyoxal-denatured RNA, but the gel background is sometimes higher; glyoxal denaturant can interfere with binding of the stain, but the gel backgrounds are often lower than with formaldehyde-denaturant systems.

As with all work with RNA, it is important to minimize RNase activity when running agarose gels by following the precautions described in Chapter 4, Section 2. Electrophoresis tanks, casting trays, and combs should be rendered free from RNases using either DEPC-treated water and hydrogen peroxide as described in *Protocol 13*.

Protocol 13

Removal of ribonuclease contamination from electrophoresis equipment

Reagents

- Detergent solution
- 3% (v/v) hydrogen peroxide
- DEPC-treated water (see Chapter 4, Section 2)

Method

1 Clean equipment with detergent solution, then rinse with distilled water and dry with ethanol.

2 Fill the gel apparatus with 3% hydrogen peroxide and soak for 10 min.

3 Rinse thoroughly with DEPC-treated water.

Note

Electrophoresis apparatus should never be directly exposed to DEPC because acrylic is not resistant to DEPC.

4.2 Preparation of RNA samples

The choice of method for sample denaturation depends largely on the final goal of the experiment and the secondary structure of the RNA. There are several procedures to choose from, the most useful of which are described below. Any sample denaturation method can be used with any of the gel buffer systems. If simply checking the integrity of cellular RNA, no sample denaturation is necessary.

4.2.1 Formamide denaturation

Formamide denaturation is suitable for almost all RNA samples and should be considered if you need to retain the biological activity of your sample (17). Gels can be cast and run in standard TAE or TBE buffer systems. If there is a significant amount of secondary structure, another denaturation method should be chosen.

Protocol 14

Formamide denaturation of RNA samples

Reagents

- RNase-free water (see Chapter 4, Section 2)
- Deionized formamide (see *Appendix 2, Protocol 4*)
- 10× MOPS buffer (200 mM MOPS, pH 7.0, 50 mM sodium acetate, 10 mM EDTA, pH 8.0)

Method

1   Bring the RNA volume up to 8 μl with RNase-free water.
2   Add 2 μl of 10× MOPS buffer and 9 μl of deionized formamide.
3   Mix thoroughly then heat at 70 °C for 10 min.
4   Chill on ice for at least 1 min before loading.

4.2.2 Formaldehyde denaturation of RNA samples

Formaldehyde denaturation works well when samples are to be recovered but it is necessary to ensure that the formaldehyde is fully removed from the RNA recovered before proceeding to subsequent studies. Some enzymatic reactions, such as *in vitro* transcription, may be problematic even after assiduous removal of the formaldehyde.

Protocol 15

Formaldehyde denaturation of RNA samples

Reagents

- RNase-free water (see Chapter 4, Section 2)
- 10× MOPS buffer (see *Protocol 14*)
- 37% (v/v) formaldehyde
- Deionized formamide (see *Appendix 2, Protocol 4*)

Method

1   Bring the RNA volume up to 6 μl with RNase-free water.
2   Add 2 μl of 10× MOPS buffer, 2 μl of 37% formaldehyde and 9 μl of deionized formamide.

3 Mix thoroughly then heat at 70 °C for 10 min.

4 Chill on ice for at least 1 min before loading.

4.2.3 Glyoxal denaturation of RNA samples

Glyoxal denatures RNA by introducing an additional ring into the guanosine residues, thus interfering with G–C base pairing. Glyoxal is a very efficient denaturant, but should not be used if samples are to be recovered. A 10 mM PIPES, 30 mM bis-Tris buffer, or a 20 mM MOPS, 5 mM sodium acetate, 1 mM EDTA, 1 mM EGTA buffer can also be used for electrophoresis instead of the phosphate buffer normally recommended (18). These alternative buffer systems do not require recirculation to prevent formation of a pH gradient.

Protocol 16

Glyoxal denaturation of RNA samples

Reagents

- RNase-free water (see Chapter 4, Section 2)

- 100 mM sodium phosphate, pH 7.0: mix 5.77 ml 1 M Na_2HPO_4 with 4.23 ml 1 M NaH_2PO_4 and bring the volume up to 100 ml with RNase-free water

- DMSO

- 6 M glyoxal, 40% (v/v) solution, deionized immediately before use. Pass the solution through a small column of mixed-bed ion exchange resin (e.g. Bio-Rad AG501-X8 or X8(D) resins) until the pH is > 5.0; large volumes can be deionized and then stored frozen in aliquots at −20 °C)

Method

1 Bring the RNA volume up to 11 μl with RNase-free water.

2 Add 4.5 μl of 100 mM sodium phosphate, pH 7.0, 22.5 μl of DMSO and 6.6 μl of deionized glyoxal.

3 Mix thoroughly then heat at 50 °C for 1 h.

4 Chill on ice for at least 1 min before loading.

4.3 Electrophoresis of RNA

The choice of an agarose free of RNase contamination is of major importance. Below is a list of general guidelines based on BioWhittaker Molecular Applications products; similar considerations apply if agaroses from a different supplier are used:

- Northern blotting requires a standard-melting-temperature agarose.

- If samples are to be recovered, a low-melting-temperature agarose can be used.

- A 1.5% or 2.0% gel made with SeaKem GTG or SeaKem Gold agaroses will work for RNA molecules of 500–10 000 nucleotides.

- For RNAs smaller than 500 nucleotides, use a 3% or 4% NuSieve 3:1 or Meta-Phor agarose gel.

- For RNA larger than 10 000 nucleotides, SeaKem Gold agarose will be a better choice for tighter bands and better resolution.

- If a low-melting-temperature agarose is required, a 1.5% or 2.0% SeaPlaque GTG gel can be used for separation of RNAs from 500–10 000 nucleotides, while a 3.0% or 4.0% NuSieve GTG gel should be used for fine resolution of RNAs smaller than 500 nucleotides. NuSieve GTG agarose is not recommended for formaldehyde/MOPS gels.

4.3.1 Electrophoresis of gels containing formaldehyde

Formaldehyde denaturant with a MOPS buffer is the most commonly used system for RNA electrophoresis. Care should be taken when handling gels containing formaldehyde as these gels are less rigid than other agarose gels.

Protocol 17

Electrophoresis of RNA in a formaldehyde gel

Equipment and reagents

Caution: Formaldehyde is a confirmed carcinogen. Whenever possible, solutions of formaldehyde should be handled in the fume hood. Gloves should be worn when handling solutions containing formaldehyde. Electrophoresis tanks should be kept covered during electrophoresis.

- Horizontal electrophoresis apparatus plus accessories (e.g. power supply, gel plate, comb)
- 10× MOPS buffer (see *Protocol 14*) prewarmed to 60°C
- Fume hood

- 37% (v/v) formaldehyde prewarmed to 60°C
- Formaldehyde loading buffer (1 mM EDTA, pH 8.0, 0.4% [w/v] bromophenol blue, 0.4% [w/v] xylene cyanol, 50% [v/v] glycerol)

Method

1. For a 1% gel, dissolve 1.0 g of agarose in 72 ml water (see *Protocol 1*). Adjust the amounts for different per cent gels.

2. Cool to 60°C in hot water bath.

3. Place in a fume hood. Without delay, add 10 ml of prewarmed 10× MOPS buffer and 5.5 ml of prewarmed 37% formaldehyde.

4. Cast the gel in fume hood, using the procedure described in *Protocol 2*.

5. Denature the RNA samples by one of the methods described in *Protocols 14–16.*[a]

6. Add 2 μl of formaldehyde loading buffer per 20 μl of sample and mix thoroughly.

7 Remove the comb and cover the surface of the gel to a depth of 1 mm with 1× MOPS buffer.

8 Load the samples and electrophorese at a maximum of 5 V/cm. Continue the electrophoresis until the bromophenol blue has travelled at least 80% of the way through the gel (see *Table 5.9* for information on dye mobility).

[a] Do not exceed 20 μg of RNA per lane as larger amounts can result in a loss of resolution.

Table 9 Dye mobilities for RNA gels

| | Running position of the dye markers (nt)[a] | | | |
| | Formaldehyde gels | | Glyoxal gels | |
Agarose (% w/v)	XC[b]	BPB[b]	XC[b]	BPB[b]
SeaKem Gold agarose				
1.0	6300	660	9500	940
1.5	2700	310	4300	520
2.0	1500	200	2300	300
SeaKem GTG and SeaKem LE agaroses				
1.0	4200	320	7200	740
1.5	1700	140	2800	370
2.0	820	60	1600	220
SeaPlaque and SeaPlaque GTG agaroses				
1.0	2400	240	4400	400
1.5	800	80	1900	180
2.0	490	30	1050	120
NuSieve 3:1 agarose				
2.0	950	70	1600	155
3.0	370	20	740	75
4.0	190	5	370	40
NuSieve GTG and MetaPhor agaroses				
2.0	Not applicable		1300	150
3.0	Not applicable		480	60
4.0	Not applicable		260	40

[a] The data relate the mobilities of the dyes to the migration rate of single-stranded RNA. Figures less than 100 nucleotides are estimated by extrapolation.

[b] Abbreviations: BPB, bromophenol blue; XC, xylene cyanol.

4.3.2 Electrophoresis of gels containing glyoxal/DMSO

Glyoxal/DMSO electrophoresis is another common technique for keeping RNA denatured. These gels should be run slower than formaldehyde ones with buffer recirculation to avoid the formation of a pH gradient. Glyoxylated RNA will give sharper bands than formaldehyde-treated RNA.

Protocol 18

Electrophoresis of RNA in a glyoxal/DMSO gel

Equipment and reagents

- 100 mM sodium phosphate, pH 7.0 (see *Protocol 16*)
- Glyoxal loading buffer (10 mM sodium phosphate, pH 7.0, 0.25% [w/v] bromophenol blue, 0.25% [w/v] xylene cyanol, 50% [v/v] glycerol)

- Horizontal electrophoresis apparatus plus accessories (e.g. power supply, gel plate, comb)
- Apparatus for recirculating the electrophoresis buffer

Method

1   For a 1% gel, dissolve 1.0 g of agarose in 100 ml 10 mM sodium phosphate, pH 7.0 (see *Protocol 1*). Adjust the amounts for different per cent gels.

2   Cool to 60 °C in hot water bath.

3   Cast the gel using the procedure described in *Protocol 2*. Gels should be cast to a thickness where the wells can accommodate 60 μl.

4   Denature the RNA samples as described in *Protocol 16*.

5   Remove the comb and cover the gel with 10 mM sodium phosphate, pH 7.0, to a depth of 1 mm.

6   Add 12 μl of glyoxal loading buffer per 45 μl of sample and mix thoroughly.

7   Load the samples and electrophorese at a maximum of 4 V/cm while the running buffer is being recirculated. Continue the electrophoresis until the bromophenol blue reaches the end of the gel (see *Table 5.9* for information on dye mobility).

Notes

(a) Do not add ethidium bromide to glyoxal gels, as this dye will react with the glyoxal.

(b) Alternative buffer systems that do not require recirculation can also be used. See Section 4.2.3.

(c) Load 0.5–1.0 μg of RNA per lane.

(d) If no recirculation apparatus is available, pause the electrophoresis every 30 min and remix the buffer in the apparatus.

4.3.3 Mobility of dyes in RNA gels

The zone of highest resolution for RNA separation in an agarose gel is in the middle third of the gel. Migration data for the common mobility dyes bromophenol blue and xylene cyanol are shown in *Table 9*. These data relate the mobility of the dyes to the migration rate of single-stranded RNA and are of assistance in determining optimal run times.

4.3.4 Detecting RNA with ethidium bromide

The procedure for detecting RNA with ethidium bromide is similar to that for DNA. Because of the reaction between ethidium bromide and glyoxal, RNA gels should be post-stained using the method described in *Protocol 3A*. Ethidium bromide does not stain RNA as efficiently as it does DNA, so ensure that sufficient RNA is loaded to see the band of interest. Visualization of poorly staining RNA is made even more difficult by the higher background fluorescence of RNA gels. In formaldehyde gels, background fluorescence is minimized with the final formaldehyde concentration in the gel at 0.6 M. Background fluorescence in glyoxal gels can be reduced by staining and destaining in 0.5 M ammonium acetate instead of water. Alternatively, RNA can be prestained during denaturation with glyoxal (18, 19), or formaldehyde (20, 21); see these references for specific details on the necessary modifications required to the denaturation and buffer systems.

References

1. Rochas, C. and Lahaye, M. (1989). *Carbohyd. Polym.* **10**, 289.
2. Himenz, P. C. (1977). *Principles of colloid and surface chemistry*. Marcel Decker, New York.
3. Adamson, A. W. (1976). *Physical chemistry of surfaces*. Wiley, New York.
4. Rees, D. A. (1972). *Biochem. J.* **126**, 257.
5. Arnott, S., Fulmer, A., Scott, W. E., Dea, E. C. M., Moorhouse, R., and Rees, D. A. (1974). *J. Mol. Biol.* **90**, 269.
6. Griess, G. A., Moreno, E. T., Easom, R. A., and Serwer, P. (1989). *Biopolymers* **28**, 1475.
7. Vandenplas, S., Wiid, I., Grobler-rabie, A., Brebner, K., Ricketts, M., Wallis, G., Bester, A., Boyd, C., and Mathew, C. (1984). *J. Med. Genet.* **21**, 164.
8. Peats, S. (1984). *Anal. Biochem.* **140**, 178.
9. Flores, N., Valles, F., Bolivar, F., and Merino, E. (1992). *Biotechniques* **13**, 203.
10. McMaster, G. K. and Carmichael, G. G. (1977). *Proc. Natl. Acad. Sci. USA* **74**, 4835.
11. Naczynski, Z. M. and Kropinski, A. M. (1993). *Biotechniques* **14**, 195.
12. Boedtker, H. (1971). *Biochim. Biophys. Acta* **240**, 448.
13. Lehrach, H., Diamond, D., Wozney, J. M., and Boedtker, H. (1977). *Biochemistry* **16**, 4743.
14. Rave, N., Orkvenjakov, R., and Boedtker, H. (1979). *Nucl. Acids Res.*, **6**, 3559.
15. Farrell, R. E. (1993). *RNA methodologies: a laboratory guide for isolation and characterization*. Academic Press, London.
16. Gruenwedel, D. W. and Davidson, N. (1966). *J. Mol. Biol.* **21**, 129.
17. Gallitelli, D. and Hull, R. (1985). *J. Virol. Method.* **11**, 141.
18. Burnett, W. V. (1997). *Biotechniques* **22**, 668.
19. Grundemann, D. and Koepsell, H. (1994). *Anal. Biochem.* **216**, 459.
20. Kroczek, R. A. (1989). *Nucl. Acids Res.* **17**, 9497.
21. Ogretmen, B., Ratajczak, H., Kats, A., Stark, B. C., and Gendel, S. M. (1993). *Biotechniques* **14**, 932.

Chapter 6
Recovery of DNA from electrophoresis gels

Paul Towner

College of Health Science, University of Sharjah, PO Box 27272, UAE

1 Introduction

During molecular biology projects it is frequently necessary to isolate DNA fragments for further manipulations. The DNA could be from many sources but would typically be from PCRs or restriction enzyme digestion of recombinant plasmids. The simplest way to separate DNA is by agarose gel electrophoresis in the presence of ethidium bromide, the required DNA bands being identified using a UV transilluminator. The quality of agarose is now so good that concentrations as high as 8% (w/v) can be cast and used to separate DNA fragments down to 30 bp with very good resolution, allowing virtually all double-stranded DNA fragments to be isolated.

Many techniques have been developed to isolate DNA from agarose; these including phenol extraction, freeze–squeeze of gel slices, digestion of the gel using agarase, electroelution onto dialysis membranes or ion exchange paper, and melting of low-gelling-point agarose followed by ion-exchange chromatography. Each of these procedures has its proponents but the methods will not be described here because they have been superseded. The overwhelmingly successful and reliable technique which has replaced these methods is based on the use of guanidinium thiocyanate and silica derivatives. The manipulations are quick to perform and result in excellent yields of high-quality DNA.

Single-stranded DNA in the form of chemically synthesized oligonucleotides can be separated and purified using polyacrylamide gels in the presence of 7 M urea. Short oligonucleotides of 15–45 bases are usually pure enough direct from the synthesizer for use in sequencing, PCR and as complementary partners in linkers. Oligonucleotides of over 50 bases used for gene construction usually require purification because traces of terminated syntheses can interfere with subsequent hybridization steps. DNA can be isolated from polyacrylamide by a crush and soak procedure in reasonable yield and the resulting DNA can be used for most purposes following precipitation with ethanol.

2 Agarose gels

High-quality electrophoresis-grade agarose suitable for the separation of DNA fragments may be obtained from many suppliers. In addition to the normal-gelling-temperature agarose, a low-melting-point (LMP) agarose is also available. LMP agarose is basically an ethylated derivative and was once well worth the extra cost because of the absence of contaminating sulphonates which would normally co-purify with the DNA. The quality of the DNA recovered from normal-gelling-temperature and LMP agarose is identical when using the method described here, and it does not matter whether Tris-borate or Tris-acetate buffer systems are used (see *Table 5.1*). There is no upper limit to the size of double-stranded DNA fragments which can be recovered from agarose gels, although large chromosomal DNA molecules, separated by PFGE, usually become sheared during the extraction. Conversely, using specifically designated proprietary kits there is no lower limit to the size of fragments which can be recovered. The techniques described in Section 2.1 give good recoveries with fragments larger than 150 bp.

2.1 Recovery of DNA from agarose gels

Several suppliers provide proprietary kits for the isolation and recovery of DNA, and their cost has become quite favourable due to the increased demand for molecular biological techniques in many laboratories. Some kits are specifically designed for use with agarose and in the majority of cases are based on the chaotropic properties of guanidinium thiocyanate which disrupts and dissolves the agarose and allows the DNA to bind to silica derivatives. The silica can be in the form of diatomaceous earth suspended in the guanidinium solution or as glass-fibre filters.

The method for recovery of DNA from agarose gels can be performed using microfuge tubes and centrifuging their contents at appropriate stages. This tends to make the procedure more time consuming and it is not as effective as using flow-through columns which can be attached to a vacuum line. As described in Chapter 3, Section 2.1, home-made columns can be prepared by piercing the base of a 0.6 ml microfuge tube with a 22 G syringe needle but it is better to use commercial microcentrifuge tube filters which comprise of a microfuge tube and a filter unit which fits inside it, such as those available from Whatman and Hybaid and Qiagen. Glass microfibre filters can also be prepared cost effectively from large diameter sheets of Whatman GF/C or GF/D filters by punching out 7 mm diameter discs using a sharp cork borer. Read the relevant part of Chapter 3, Section 2.1 for the background information on these systems.

The first step in the procedure for recovery of DNA from an agarose gel is sol-ubilization of the gel using a guanidinium thiocyanate buffer. This is described in *Protocol 1*.

The DNA in the solubilized agarose can be isolated by mixing with diato-maceous earth (*Protocol 2*) or by passing the solution through a glass-fibre filter (*Protocol 3*). In the latter procedure the agarose is removed from the filter matrix with the solubilization buffer and the guanidinium thiocyanate is then washed way with a high-salt ethanol solution. This in turn is removed from the column by centrifugation and the DNA eluted from the matrix.

Protocol 1

Solubilization of an agarose gel

Reagents

- Gel solubilization buffer (dissolve 472 g guanidinium thiocyanate (Sigma G9277) in about 0.5 l of water; add 50 ml 1 M Tris-HCl, pH 7.5, and 40 ml 0.5 M EDTA, pH 8.0, and make up to 1 l; store in an airtight screw-topped light-proof container)[a]

Method

1 Identify the DNA fragment of interest in the gel and carefully isolate the block of agarose in which it is contained. Using a scalpel or single-edge razor blade, being careful not to damage the UV transilluminator, remove as much waste agarose as possible. **Caution: Wear suitable eye and skin protection when manipulating the gel on the UV transilluminator.**

2 Transfer the gel slice to a preweighed microfuge tube, or for larger gel pieces a 5 or 20 ml universal tube. Note the gel weight and, if possible, slice it into very small pieces. Add 5 volumes of gel solubilization buffer. This is suitable for agarose in the 0.5–2.5% range; higher concentrations require more dissolving buffer.

3 Place the tube in a heated water bath at 42–55 °C. Finely sliced agarose will dissolve within 1–2 min, but larger blocks will take progressively longer and will benefit from occasional gentle shaking. The temperature is not critical since the gel will dissolve if left at room temperature.

4 Ensure the agarose has dissolved entirely.

[a] The loose hygroscopic crystals of guanidinium thiocyanate occupy around 600 ml. They dissolve endothermically so it is expedient to use hot water.

Protocol 2

Purification of DNA from dissolved agarose using DNA-binding resin

Equipment and reagents

- Diatomaceous earth (Sigma D-5384)
- Gel solubilization buffer (see *Protocol 1*)
- Filter units (Whatman, Hybaid or Qiagen)
- Ethanol wash solution: 50 mM Tris-HCl, pH 7.5, 10 mM EDTA, pH 8.0, 0.2 M NaCl, 50% (v/v) ethanol

Protocol 2 continued

A. Preparation of the DNA-binding resin

1 Place 10 g of diatomaceous earth in a 2-l transparent vessel and stir in 2 l of water to resuspend the powder. Allow to settle for 1 h and pour away the supernatant containing any material which remains in suspension.

2 Resuspend the diatomaceous earth in a further 2 l of water and again allow to settle for 1 h before pouring away the supernatant.

3 Prepare solubilization buffer as in *Protocol 1* but restrict the volume to 900 ml and use this to resuspend the diatomaceous earth before making up to 1 l.

4 Store in an airtight screw-topped light-proof container.

B. Purification of DNA

1 Resuspend the DNA-binding resin and add to the dissolved gel an amount of resin that is equal to the original gel volume (e.g. 0.1 ml/100 mg agarose gel).

2 Mix gently for several seconds to allow the DNA to bind to the resin.

3 Transfer the tube contents to a filter unit attached to a vacuum line and allow the liquid to drain from the filter.

4 Pipette 2 ml gel dissolution buffer into the filter unit, ensuring that it displaces any trace of liquid remaining on the sides.

5 Pipette 2 ml ethanol wash solution into the tube, again ensuring that all traces of liquid in the tube are displaced.

6 Transfer the filter unit into a microfuge tube and centrifuge for 2 min at 15 000 g to remove all traces of liquid.

7 Transfer the filter unit to a fresh microfuge tube and carefully pipette 20–50 µl water onto the bed of resin and allow it to soak in.

8 Centrifuge for 2 min at 15 000 g to collect the liquid containing the DNA.

Notes

(a) The initial steps of *Protocol 2A* are designed to remove very small particles of silica which would otherwise block the holes in membranes and sintered discs. Consequently, the concentration of resin is actually less than 10 g/l, but the concentration is not critical. Stocks of up to 20 g/l can be prepared.

(b) The amount of DNA-binding resin required should be based on the amount of DNA in the gel. One millilitre of a 10 g/l resin solution can bind in excess of 5 µg DNA. The DNA binds to the resin almost instantaneously but may require a few minutes if the volume of the dissolved gel is more than a few millilitres. With large volumes, the resin can be left to settle after the DNA has bound: this allows the supernatant to be removed so that excessively large volumes are not pipetted to the column.

(c) It is important that the solution containing agarose is removed from the column by washing with dissolution buffer, otherwise traces of agarose will cause the DNA to bind irreversibly to the resin at the ethanol wash step.

(d) The filter unit must be centrifuged to remove all traces of salt and ethanol because, in most cases, the DNA will be eluted in a very small volume of water. The elution step may be repeated if required, or larger volumes of eluent used in proportion to the amount of DNA. If desired, 1 mM Tris-EDTA can be substituted for water. It is unnecessary to heat the eluent. Have all reagents to hand so that air is not allowed to be sucked through the resin bed between the different wash steps. If traces of resin have leaked through the sinter it is important to transfer the DNA solution to a fresh tube, avoiding the pellet and any traces of resin, otherwise any modification enzymes that are subsequently added to the solution will bind to the resin and become inactivated. If filter units are not employed, all of the steps can be performed using a microfuge tube and the supernatants discarded by aspiration after brief centrifugation (10–20 s). Each of the solvent exchange steps should be duplicated to ensure that there is no carry-over of unwanted solutes.

Protocol 3

Purification of DNA from dissolved agarose using glass fibre filters

Equipment and reagents

- Glass fibre filter discs, 7 mm diameter (Whatman)
- Gel solubilization buffer (see *Protocol 1*)
- Ethanol wash solution (see *Protocol 2*)

Method

1. Prepare a filter column with two to ten 7-mm diameter glass-fibre filter discs. Attach to a water-jet pump at a very low flow rate.

2. Pipette the dissolved agarose sample onto the filter column such that 1 ml of the solution requires 1 min to flow through the column.

3. Pipette 1.0 ml gel solubilization buffer into the filter so that all traces of agarose are removed.

4. Pipette 1 ml ethanol wash solution into the filter column to remove all traces of solubilization buffer.

5. Transfer the filter column to a microfuge tube and centrifuge at 15 000 *g* for 2 min.

6. Place the filter column in a fresh microfuge tube and pipette 20–50 μl water onto the centre of the glass-fibre filter and allow it to soak in.

7. Centrifuge for 2 min at 15 000 *g* to collect the liquid containing the DNA.

Notes

(a) Unlike the use of diatomaceous earth, the DNA has the opportunity to bind to the glass-fibre filter only as it flows through the column. If the flow rate of the liquid is too high, substantial losses of DNA can result. After the initial binding step the water-jet flow rate can be increased.

(b) In the absence of filter units, glass-fibre filters can be packed into the base of a 0.6 ml microfuge tube which has been pierced at its apex.

2.2 Limitations on DNA recovery from agarose

There are many documented methods used for the isolation of DNA from agarose but few fulfil the stringent test of obtaining clonable DNA fragments without the need of further purification. The technique described here, using guanidinium thiocyanate and silica derivatives for DNA isolation, is derived from the method outlined by Carter and Milton (1) for purification of plasmid DNA and has simply been applied to DNA embedded in agarose. The only drawback to the method is that with the reagents described the recovery of DNA below 150 bp cannot be assured. By lowering the pH of the solubilized gel, increased binding of small DNA fragments to the silica matrix is achieved. For example, using a proprietary kit from Qiagen with a buffer designed specifically for small DNA fragments it is possible to recover DNA fragments of 50 bp in high yield. The range of kits is extensive and many are optimized for specific types of DNA. Furthermore, by purchasing such a kit, a stock of recycled flow-through columns will be available for use in the protocols described. This will save on resources but it should be emphasized that the real advantage of recycling is when the flow-through columns are also used for the many plasmid mini-preparations that every molecular biologist carries out (Chapter 3, Section 2.3). It is also very important to note that guanidinium thiocyanate is poisonous and must never be mixed with strong acids or bleaching agents as toxic gases result.

3 Polyacrylamide gels

The most effective separation method for single-stranded DNA is electrophoresis through polyacrylamide in the presence of 7 M urea. In these denaturing conditions the single-stranded molecules have no opportunity to base pair and hence separate as discreet bands with a resolution of a single nucleotide. Denaturing polyacrylamide gels are generally used at 8–20% (w/v) acrylamide in Tris-borate buffers to separate synthetic oligonucleotides of 10–125 bases. Lower concentrations of acrylamide can be used to separate single-stranded nucleic acids several-hundred bases long. Occasionally, non-denaturing polyacrylamide gels of 2–10% are used which can separate double-stranded DNA of 20–1000 bp, but these have now largely given way to high-resolution agarose gel electrophoresis (see Chapter 5, Section 3). Polyacrylamide gels are electrophoresed vertically between two glass plates and may be cast in a variety of sizes. Analytical gels of approximately 50×80 mm can be used to assess the synthesis quality of oligonucleotides which can be detected within the gel by silver staining or autoradiography of material that has been end-labelled (using polynucleotide kinase) with radioactive phosphorus in the form of phosphate from appropriately labelled triphosphonucleotides. Preparative gels are about 200×200 mm with a gel thickness of 2.0 mm. A gel lane of 150 mm can accommodate all of the material obtained from a 0.2 μmol oligonucleotide synthesis.

After electrophoresis, the bands of interest can be visualized by UV shadowing directly on the electrophoresis plate if it is fluorescent. Otherwise, the gel should be transferred to UV transparent cling film and irradiated with UV such that a fluorescent TLC plate placed directly under and close to the gel will show the position of the DNA as a shadow on the fluorescent plate because of its UV absorption properties. This technique is only effective with relatively large amounts of material, smaller quantities of DNA only being visible after the gel has been briefly covered in ethidium bromide (1 mg/ml) and supported upside-down above a transilluminator. This is easily done if the periphery of the gel is kept in contact with the glass plate by covering in cling film. The separated bands of DNA can be recorded by copying their positions on the rear of the electrophoresis plate with a pen and, by inverting the plate, slices of gel containing the DNA can be excised.

3.1 Recovery of DNA from polyacrylamide gels

It is unnecessary to purify oligonucleotides prepared by chemical synthesis unless they fail in the purpose for which they were intended. Synthesis machines are now so good that any sequence up to 30 nucleotides in length can usually be obtained as a pure full-length product. With much longer oligonucleotides problems can arise because the proportion of failed syntheses increases and the presence of these failed products can interfere with subsequent uses of the oligonucleotide preparation. In gene synthesis procedures, for example, two 120-mers could be synthesized which are complementary over 20 bp at their 3'-ends. By hybridizing the two oligonucleotides together and extending them with a DNA polymerase a 200-bp DNA fragment should result. However, it is more than likely that the 200-bp fragment will only be obtained if the full-length oligonucleotides have been electrophoresed through and isolated from a polyacrylamide gel. After isolation from the gel the oligonucleotide can be precipitated with ethanol (see Chapter 3, *Protocol 8*) and used in further manipulations. The method described in *Protocol 4* is also suitable for the recovery of much longer strands of nucleic acid and is routinely used to isolate RNA molecules of several hundred nucleotides.

Protocol 4

Recovery of DNA from polyacrylamide gels

Reagents

- TE, pH 7.4 (10 mM Tris-HCl, pH 7.4, 1 mM EDTA, pH 8.0)

- 0.3 M sodium acetate, pH 6.8 (prepared by adding glacial acetic acid to 0.3 M sodium acetate until pH 6.8 is obtained) + 1 mM EDTA

Method

1 Slice the band of interest from the gel and place it in a 20 ml universal tube if the material is from a 2-mm thick gel. Smaller gel slices from 0.34-mm thick sequencing gels can be placed in a microfuge tube.

Protocol 4 continued

2 Oligonucleotide DNA from 2-mm thick gels is extracted by adding sufficient 0.3 M sodium acetate, pH 6.8, 1 mM EDTA to cover the gel in the tube, which is left to mix on a table shaker for 12–18 h at ambient temperature.

3 Polynucleotide samples from 0.35 mm gels of lower acrylamide concentration are extracted by adding buffer as in step 2 to just cover the slice and storing at 4 °C for 3–6 h.

4 Remove and save the buffer in which the gel was bathed, avoiding any particles of polyacrylamide, and transfer to a suitable centrifuge tube. Gel samples in microfuge tubes should be re-extracted with fresh buffer. The transferred solutions containing the DNA should be centrifuged at 15 000 g for 10 min to remove any particles of polyacrylamide.

5 Add 3 volumes of cold absolute ethanol to oligonucleotide samples, and 2.5 volumes to polynucleotides, incubate at −20 °C for 30 min and then centrifuge at 15 000 g and 0 °C for 30 min.

6 Discard the supernatant and rinse the tube contents carefully with 70% (v/v) ethanol.

7 Discard the supernatant and dry the tube *in vacuo*.

8 Take up the DNA sample in TE, pH 7.4.

9 Assess the yield by taking an absorbance spectrum as described in Chapter 3 *Protocol 9*.

Notes

(a) Extraction of nucleic acid from polyacrylamide is best done on whole gel slices. If the slices are fragmented the yield of DNA does not improve, but problems ensue because particles of polyacrylamide contaminate the DNA when it is being precipitated.

(b) Derivatives of *bis*-acrylamide are available which can be used to prepare polyacrylamide substitutes that can be treated as if they were agarose gels. This is because they dissolve in mild reducing conditions and even more quickly in the presence of guanidinium thiocyanate, which allows the oligonucleotides to bind to silica resins. These acrylamide derivatives have not been widely used simply because normal polyacrylamide is well-tried and tested and recovery of nucleic acid by the method described here is more than adequate for most purposes.

Acknowledgement

I am indebted to Dr Nicholas Lea for suggesting the use of glass fibre filter discs and demonstrating the effectiveness of these protocols.

Reference

1. Carter, M. J. and Milton, I. D. (1993). *Nucl. Acids Res.* **21**, 1044.

Chapter 7

Construction of recombinant DNA molecules

Richard Powell
Department of Microbiology, National University of Ireland, Galway, Ireland

Frank Gannon
European Molecular Biology Laboratory, Postfach 10.2209, Meyerhofstrasse 1, D–69012 Heidelberg, Germany

1 Introduction

Although the central role of DNA in all living forms was well accepted by the 1950s it was very difficult to study it directly at that time (1, 2). The information obtained about DNA came mostly from the interpretation of classical genetic experiments. Most of these were performed using microorganisms whose relative simplicity and ease of manipulation made them the organisms of choice. The difficulty of obtaining a detailed understanding of a bacterial virus or phage such as λ in which there are a mere 50 000 nucleotides (approximately) made the wish to understand the structure and function of higher organisms in which there are almost 10^5 times more nucleotides seem unattainable. Then, in the mid-1970s, a technological revolution occurred which has allowed this great chasm in accessible information to be bridged. As in all revolutions, developments in many unrelated areas came together in a way which magnified the individual components into a powerful new ensemble.

Research from many unrelated and frequently obscure areas of biochemistry, microbiology, chemistry, virology, and immunology provided the elements we now recognize as recombinant DNA technology or genetic engineering. The essential result of this collage of techniques is that it is now possible to purify and amplify DNA fragments from any biological material. Direct analysis of DNA sequences is now a daily event in laboratories in all biological disciplines worldwide. The rapid growth of biotechnology as an industrial sector has been fuelled by the ability to express the proteins encoded by the isolated DNA fragments in host cells that grow readily in bioreactors. Our understanding of the detailed interactions which combine to allow the controlled expression of genes has contributed to areas as diverse as medicine and agriculture and molecular biologists are recognized as having valuable perspectives to bring to bear on many previously recalcitrant problems.

After two decades of practice, the early steps in a genetic engineering experiment have been reduced to a simple equation:

$$\text{vector} + \text{foreign DNA} + \text{host cell} = \text{gene library}$$

It is frequently a surprise to those entering the field that the experimental procedures involved, which have been analysed to an extent that they no longer warrant mention, are far from trouble-free. It should be reiterated that the skills and care of a biochemist are required to purify DNA free from all biological and organic factors which can interfere with subsequent steps, that the enzymes used are delicate tools which demand correct handling, and that the biological hosts which serve as recipients for the foreign DNA require the knowledge and experience of a trained microbiologist. When all of these considerations are met then the construction of recombinant DNA molecules can and does work efficiently.

The demand for such molecules has given rise to a very large and supportive service industry. Every component for a recombinant DNA experiment can be purchased. Kits are available in which the skill required to obtain the molecule of interest is reduced to the ability to follow some very simple pipetting steps. Those who might find that too demanding can purchase ready-made recombinant DNA libraries. However, most laboratories, even if they purchase ready made DNA clones, need to propagate, alter, sequence or generate new constructs from the DNA fragments. As such the A-B-C of digesting DNA, ligating it and using it in transformations has to be learned and appreciated by all. By becoming proficient in these skills the useful results which come from complex methods become readily attainable and reproducible.

2 Restriction enzyme digestions

In all aspects of genetic engineering, the strategy employed to reach the goal of the project comes from a sequence of major choices. The starting DNA is probably dictated by the sequence which is sought but the manner in which it is digested, the vector which is used and the host in which it is inserted are all *E coli* interrelated choices. Since the essential result from a genetic engineering experiment is the purification of a fragment of DNA which is more simple to analyse than that available in the animal, plant or microbial cell, the first step in processing purified DNA (see Chapter 3) is the generation of smaller and more manageable fragments. Almost without exception, this is now achieved by use of specific microbial nucleases called restriction enzymes (3).

The substrate nucleotide sequences at which each of several-hundred restriction enzymes digest DNA are known (see *Appendix 3*). An examination of these will show that they are characterized by being palindromic. The manner in which these enzymes digest the DNA—leaving a 5′ protruding sequence, a 3′ protruding sequence, or blunt (flush) ends—is also known. Because of the palindromic nature of the restriction enzyme sites, fragments generated have complementary or 'sticky' ends. Ultimately, these termini will be linked (ligated) to vector DNA (see Section 4.1) with homologous or compatible termini. Appropriate complementary termini can also be generated by two different enzymes. For example, digestion by *Bam*HI and by *Sau*3AI both yield the complementary sequence 5'-GATC-3'. Methods for routine restriction enzyme digestion by one or more restriction enzymes and the subsequent step of denaturing the restriction enzyme are shown in *Protocols 1–3*.

Protocol 1

Restriction enzyme digestion

Reagents

* Restriction enzyme plus appropriate buffer

Method

1 Make up a 100 μl reaction mixture[a] by adding the reagents in order into a sterile 1.5 ml microfuge tube: 10 μl 10× restriction enzyme buffer[b], x μl DNA and y μl water. The value of x will depend on the concentration of your DNA solution.[c] The value of y should bring the volume up to 99 μl.

2 Add 1 μl of the restriction enzyme[d] and mix gently with the pipette tip.

3 Incubate in a water bath at the appropriate temperature for the desired length of time. The temperature and time depends on the enzyme and will be stated in the product guide provided by the supplier.

4 Store the reaction on ice while analysing the digestion by gel electrophoresis.

[a] When used for analytical purposes only, the reaction can be carried out in a volume of 20 μl.

[b] Use the buffer provided by the suppliers of the restriction enzyme, or make a buffer according to their instructions.

[c] Although concentrations of DNA as low as 10 ng can be visualized on an agarose gel, it is preferable to use 100–200 ng per digestion to facilitate easy detection of small DNA fragments.

[d] Add about 1 unit of enzyme per μg of DNA. If necessary dilute the enzyme stock with the storage buffer described in the supplier's product guide. It is acceptable to add more than 1 unit/μg of DNA, but large excesses should be avoided.

Protocol 2

Sequential digestion by two restriction enzymes

Equipment and reagents

* Restriction enzymes plus appropriate buffers

* Gel electrophoresis apparatus

Method

1 Set up a reaction mixture as described in *Protocol 1*. This reaction mixture should be appropriate for the restriction enzyme requiring the lower salt concentration of the two enzymes to be used. If the reaction buffers of both enzymes are similar, the DNA may simply be incubated in the presence of both enzymes.

2 Incubate at the correct temperature for the desired length of time and place the reaction on ice for the time required for gel electrophoretic analysis.

3 Electrophorese a small sample of the reaction (see Chapter 5, Section 2) and check for total digestion of the DNA. If the digestion is not total, reincubate the reaction mixture for a further period of time. Alternatively, add a further aliquot of enzyme and re-incubate.

4 After electrophoretic analysis showing completion of digestion by the first restriction enzyme, change the salt concentration to that appropriate for the second restriction enzyme (see ref. 4) by adding a small aliquot (1–5 µl) of 5 M NaCl to the reaction mixture. Add the second enzyme and allow this reaction to progress before monitoring for correct double digestion by gel electrophoresis.

Notes

(a) This method is adequate for many double digestions. However, in some cases the reaction conditions of the two enzymes may not be compatible with simple alteration of the salt concentration. Also, the first enzyme may lose its DNA specificity and begin to cleave the DNA in an unexpected manner while under the reaction conditions for the second enzyme. In these cases the first enzyme may be completely inactivated after reaction by incubation at 65 °C for 10 min before continuing with the second reaction. However, many enzymes are not heat-labile (4) and the DNA must therefore be purified from the reaction mixture by phenol extraction and ethanol precipitation (*Protocol 3*) before continuing with the second restriction enzyme.

(b) Complete double digestion of DNA by two enzymes recognizing restriction sites adjacent or very close to one another (as in a polylinker multicloning site) may not be distinguishable by simple gel electrophoresis from DNA cleaved by only one of the enzymes: both fragments may appear to be of similar size. A simple ligation reaction (*Protocol 9*) after DNA purification (*Protocol 3*) may be required to ensure correct double digestion. If the product of the double digestion is a DNA fragment with non-complementary termini, unlike a DNA fragment cleaved by one restriction enzyme, it cannot ligate to itself to form circular molecules. These differences are easily analysed by gel electrophoresis.

Protocol 3

Purification of DNA by phenol extraction and ethanol precipitation

Reagents

- Phenol (prepared as described in Chapter 3 *Protocol 6*)
- TE buffer, pH 8.0 (10 mM Tris-HCl, pH 8.0, 1 mM EDTA, pH 8.0)

- 24:1 (v/v) chloroform–isoamyl alcohol (prepared as described in Chapter 3 *Protocol 6*)

- 3 M potassium acetate, pH 5.5, prepared by adding glacial acetic acid to 3 M potassium acetate until this pH is obtained

Method

Caution: phenol is one of the most hazardous chemicals used in molecular biology (see Chapter 1, Section 3.3).

1 Add an equal volume of phenol to the reaction mixture and vortex gently.

2 Separate the aqueous phase which contains the DNA from the organic phase by centrifugation in the microfuge, at 2000 g for 5 min or at 8 000 g for 1 min.

3 Remove the aqueous phase with care into a fresh microfuge tube and add an equal amount of 24:1 (v/v) chloroform–isoamyl alcohol.

4 In order to precipitate the DNA, add a 0.1 volume of 3 M sodium acetate, pH 5.5, to the aqueous phase and then 2 volumes of absolute ethanol. Incubate at −20°C overnight or for much shorter periods at −80°C (20–30 min).

5 Pellet the precipitated DNA by centrifugation in the microfuge at 8 000 g for 5–15 min. Remove the ethanol with care and dry the pellet in a desiccator or 50°C oven for 5 min. An extra wash with 70% (v/v) ethanol may be included to remove excess salt from the pellet. The dried DNA may be resuspended in sterile TE, pH 8.0, or water, and stored at 4°C for further manipulation.

6 This procedure denatures and removes contaminating protein from a DNA sample. A second useful method is drop dialysis, which can remove salt, SDS, and even enzyme inhibitors. As such, it can be used with the majority of the methods in this chapter involving DNA purifications before or after enzymatic reactions:

   (a) Gently place a drop dialysis filter (Millipore VSWP 02500), floating shiny-side up, on 10–20 ml dialysis buffer (TE, pH 8.0, or water) in a Petri dish.

   (b) Pipette the DNA sample (10–100 µl) on to the filter.

   (c) Allow to dialyse for 1–2 h before removal for further analysis.

The target site for a restriction enzyme may be 4, 6 or more nucleotides in length (3, 4). This fact dictates the size of the average DNA fragment which results from complete digestion. The probability of a restriction enzyme site occurring for an enzyme with a 4-nucleotide target site is 1 in 256 (there is a 1 in 4 chance of a given nucleotide occurring at any site, so the chances of a correct 4-nucleotide sequence occurring therefore is $1/4 \times 1/4 \times 1/4 \times 1/4 = 1/256$). This means that the average size of the DNA fragments which result from a complete digestion of DNA by a restriction enzyme with a 4-nucleotide target site is 256 bp.

Fragments of that size are too small for most purposes. They are significantly smaller than the average prokaryotic gene (if one assumes that an average protein contains approximately 500 amino acids, which are encoded by 1500 nucleotides) and even more so for eukaryotic genes, most of which contain introns.

Expressed in another manner, a library with fragments of that size would require approximately 10^7 constituents to include a total eukaryotic genome and this would complicate the act of screening the library for the sequence of interest.

Finally, although plasmid vectors do not have a lower limit for the size of the fragment that can be integrated, λ and cosmid vectors do. For all of those reasons, total digestions by restriction enzymes which recognize 4-nucleotide sites are not used for the generation of genomic libraries. Two alternatives can be considered: either the use of restriction enzymes which digest the DNA less frequently (at 6- or 8-nucleotide recognition sites) or the use of incomplete or partial digestions with a frequent-cutting enzyme. The former method has the double disadvantage that although the average size might be acceptable, fragments at the extremes of the spread will either be too small or too large and the location of the restriction enzyme site might be inopportune if it occurred within the gene of interest, For these reasons, partial digestions of the DNA one wishes to use in the generation of the library are the usual if not universal choice.

Because of the frequency of occurrence of 4-nucleotide target sites the fragments incorporated into the vectors as a result of a partial digestion will closely approximate a random collection which includes the total target genome. Controlled partial digestions can be obtained by varying the amount of enzyme, the time of incubation or the conditions of the reaction from suboptimal to optimal. In all cases it is important to recall that the enzyme used is a sensitive reagent. When in very dilute solution it may be unstable and hence one-tenth of a unit of enzyme may not have one-tenth of the activity of 1 unit. Similarly, the activity of the enzyme may not be maintained without diminution over a long period. For these reasons, it is normal to carry out a scaled-down preliminary experiment to establish the optimum conditions. This should replicate as far as is possible the final full-scale experiment as regards units of enzyme/μg of DNA and the concentration of enzyme relative to the volume of the reaction.

The enzyme used will depend on the requirements of the experiment. The restriction enzymes used for partial digestions are chosen because of the compatibility of the termini which they generate within the integration sites of the vectors. For this reason Sau3AI (recognition sequence 5′-GATC-3′), for example, is frequently selected for this step as the fragments which result from the digestion can be linked to BamHI-digested (5′-GGATCC-3′) vectors.

The procedure for a partial digestion of the target DNA is shown in *Protocol 4*.

Protocol 4

Establishment of correct partial digestion conditions

Reagents

- Restriction enzyme plus appropriate buffer
- 200 mM EDTA, pH 8.0

Method

1   Prepare a 100 μl reaction mixture (*Protocol 1*) optimal for the specific restriction enzyme and containing 10 μg of genomic DNA.

2   Dispense 20 μl of the mixture into microfuge tube 1 and 10 μl into the remaining tubes 2–9. Place all tubes on ice.

3   Add 1 μl (= 1 unit) of restriction enzyme to tube 1, mix gently and place on ice.

4   With a fresh tip, pipette 10 μl from tube 1 into tube 2, mix gently and replace on ice. Perform similar serial dilutions throughout the assay tubes, thereby diluting the enzyme by 50% between each tube, keeping the enzyme–DNA mixture on ice.

5   Incubate at 37 °C for 1 h before placing the tubes on ice.

6   Add EDTA, pH 8.0, to 20 mM to inactivate the enzyme.

7   Electrophorese each sample alongside appropriate DNA size markers (see Chapter 5, Section 2). Visualize the DNA with a UV source and analyse the DNA digestion pattern in each gel track compared with the DNA size markers, to determine which enzyme concentration produces the maximum amount of DNA fragments of the desired size.

Note

The addition of spermidine can enhance restriction enzyme digestion of impure DNA. Add 0.1 M spermidine to a final concentration of 4 mM and incubate the DNA mixture at 37 °C for a few minutes (spermidine may precipitate long DNA). Then add the restriction enzyme after ensuring no DNA precipitation has occurred.

Although *Protocol 4* is simple and requires no unusual skills, one often finds that the digestion is unsatisfactory. In such circumstances a number of control experiments are recommended:

(1) Check that the enzyme and the buffer used are functioning correctly by using plasmid DNA which is known to be correctly digested by a different restriction enzyme.

(2) Distinguish between a problem with the target DNA and with the solution in which the DNA is present. This is done by adding plasmid DNA which can be digested to the solution of target DNA and carrying out the restriction enzyme reaction. If the 'clean' DNA is digested it shows that the target DNA is the source of the problem. If it is not digested, then the first priority is to remove the phenol, polysaccharide, salt or unknown inhibitor of the restriction enzyme. This can be done by a variety of methods including drop dialysis (see *Protocol 3*, step 6), binding to silica derivatives (see Chapter 6, Section 2.1) or dialysis. Simply repeating an ethanol precipitation step may not be fruitful if this had been included in the preparation of the target DNA because many inhibiting compounds are precipitated by ethanol.

(3) If the target DNA seems to be the cause of the problem, then further DNA preparations may have to be performed.

When the restriction enzyme digestion appears to be correct, the full-scale experiment can be performed. Because of the difficulty of correctly sampling small volumes from viscous solutions of undigested DNA, it is seldom that full-scale reactions require the same amount of enzyme or time. However, it is not possible to give an accurate indication of the precise conditions to use because this will ultimately depend on factors such as the stability of the restriction enzyme and the cleanliness of the DNA preparation. It will also depend on the subsequent uses intended for the DNA. For example, a cosmid library will require DNA which is, on average, very long (> 40 kb) and any overdigestion is a disadvantage.

After the digestion of the target DNA, an aliquot is used to analyse its size by electrophoresis on an agarose gel. If it is unsatisfactory, further enzyme may be added. If persistent difficulties are encountered the addition of 4 mM spermidine to the sample may prove beneficial.

The analysis of the restriction enzyme digestions is performed in agarose gels with DNA fragments of known sizes as size markers. The methods involved have been described in Chapter 5, Section 2. With partial digestions, a spread of DNA sizes is obtained. It is important to remember that 1 mole of a fragment of size 20 000 bp will give a signal on visualization with ethidium bromide which will be 10 times as intense as that of a fragment of size 2000 bp. Ultimately, it will be the number of molecules of DNA that are present in a ligation which will dictate the success of the experiment. It is advisable therefore to eliminate DNA fragments which are too small prior to further manipulations. Some vector systems make this step seem superfluous (e.g. λ systems which are non-viable with small DNA fragments; see Chapter 9, Section 4). However, the prior fractionation of DNA, usually by sucrose gradients (*Protocol 5*) after phenol extraction, has the twin benefits of size fractionation and elimination of traces of organic solvents and is recommended.

Protocol 5

Large-scale preparation and purification of partially-digested DNA

Equipment and reagents

- Restriction enzyme plus appropriate buffer
- Fraction collector
- Gel electrophoresis equipment
- 10–40 (w/v) sucrose gradient prepared in a suitable centrifuge tube (e.g. for the Beckman SW-40 rotor). These gradients

are readily achieved by freezing and thawing twice a 25% (w/v) sucrose solution in 1 M NaCl, 20 mM Tris-HCl, pH 8.0, 5 mM EDTA in a swing-out ultracentrifuge tube or by using gradient-forming devices

- Ultracentrifuge

Method

1   Selecting the optimal conditions established for correct partial digestion (*Protocol* 4), prepare a reaction mixture containing 50–100 µg of genomic DNA. Scale up the reaction mixture volume so that the reaction time, temperature and DNA concentration are identical to those established.

2   Add half the amount of restriction enzyme to that established for correct partial digestion. Few scale-up experiments produce results that are identical to pilot studies and the margin of error with this method favours underdigestion of the DNA by the enzyme and not useless over-digestion.

3   After incubation for the correct period, place the reaction on ice and analyse a small sample by gel electrophoresis. If the sample is underdigested, replace at the correct temperature for a further 15 min before repeating the electrophoresis. Addition of more enzyme at this stage usually results in overdigestion.

4   Purify the digested DNA by phenol extraction and ethanol precipitation (*Protocol* 3).

5   Apply to a 10–40% (w/v) sucrose gradient and centrifuge at 120 000 g for 24 h at 15 °C in a Beckman SW40 rotor.

6   Collect fractions and analyse samples by gel electrophoresis to determine which fraction contains DNA of the correct size or within correct size limits for the cloning experiments.

7   Purify selected DNA by dialysis against sterile water overnight to remove the sucrose and concentrate to 0.1–0.5 µg/µl by ethanol precipitation (*Protocol* 3, steps 4–5).

After the sucrose gradient, fractions are collected and are again analysed on agarose gels. Presumably, because of differences in the sample buffers, there is a tendency to overestimate the size of the DNA in these gels. Bearing in mind that small molecules tend to cause problems in molecular cloning experiments it is again advisable to have size markers appropriate for the fragment size sought and to choose fractions that appear to be slightly too large.

3 Preparation of vectors for molecular cloning experiments

The essential requirement of a vector is that it can replicate in the host organism. Phage λ, cosmids, and plasmids fulfil this basic requirement and are widely used in cloning experiments (see Chapter 9). Plasmids are extrachromosomal elements which, when used as cloning vectors, typically carry resistance to an antibiotic and are present as multiple copies in the host organism. Plasmids are most frequently used to subclone large fragments or to generate small libraries such as would be required for microorganisms, viruses and some cDNA banks.

Genomic libraries for eukaryotic organisms are generally prepared with phage λ as vector. Only the salient features of λ when used as a vector will be noted here (for a more complete discussion see Chapters 2 and 9). Phage λ can function in a lytic or lysogenic (integrated into the DNA of *Escherichia coli*) mode. For molecular cloning, only the lytic functions are required as these ensure a high copy number of the phage. Nature has conveniently arranged matters such that the lysogenic functions, which are unnecessary for λ as a vector in the service of a molecular biologist, are grouped together in the middle of the phage. This means that digestion by appropriate restriction enzymes can allow separation of the 'arms' of λ from the central 'stuffer' fragments. In the generation of genomic banks this fragment is replaced by the foreign DNA and all of the lytic functions required for replication and infection by the phage are retained. The initial transfer of the phage arms linked to foreign DNA into the host cell is effected by regenerating *in vitro* phage particles by the addition of λ head and tail proteins. This procedure of 'packaging' DNA is described in Chapter 8, Section 2.3.

The length of the DNA which can be enclosed in the phage head cannot be greater than approximately 47 000 bp. Similarly, phages which contain less than approximately 32 000 bp are also unstable. This places constraints on the size of insert which can be carried when phage λ is the vector. The packaging of λ DNA *in vitro* depends on the initial binding of capsid proteins to the *cos* sites of the phage. These must be presented in a linear array with the distance between the *cos* sites respecting the size limitations outlined above. This is achieved by the ligation of the arms of λ and the target DNA at a high concentration to form concatamers (see Section 4.1). The packaged DNA is transferred to *E. coli* with very high efficiency. This fact, allied with the ease of storage of λ genomic libraries and the large number of λ plaques which can be analysed on a single agar plate, make λ genomic libraries particularly popular.

3.1 Phage λ

When λ is used as a cloning vector the details of the strategy used depends on the particular vector. With λgtWES, for example, digestion by *Eco*RI followed directly by a size separation or with a second digestion with an enzyme (*Xba*I) which has a site only in the stuffer fragment prior to size separation yields arms of λ which will accept foreign DNA.

Protocol 6

Preparation and purification of λgtWES arms

Equipment and Reagents

- λgtWES DNA
- *Eco*RI restriction enzyme plus appropriate buffer
- 10–40 (w/v) sucrose gradient prepared in a suitable centrifuge tube (see *Protocol 5*)
- Ultracentrifuge
- Fraction collector

Method

1 Carry out a total *Eco*RI digestion of 50–100 μg λgtWES DNA (*Protocol 1*).

2 Phenol extract and ethanol precipitate the DNA (*Protocol 3*).

3 Add $MgCl_2$ to a final concentration of 10 mM to the resuspended DNA and incubate at 42 °C for 1 h. This promotes annealing of the complementary *cos* sites thereby joining the λgtWES arms.

4 Apply the DNA on to a 10–40% (w/v) sucrose gradient.

5 Centrifuge at 120 000 **g** for 24 h at 15 °C in a Beckman SW40 rotor.

6 Collect 0.5–1.0 ml fractions carefully and analyse a sample from each fraction by gel electrophoresis to determine which contain purified λ arms with no contaminating λ stuffer fragments.

7 Dialyse the appropriate fractions against sterile water overnight to remove the sucrose.

8 Concentrate the DNA by ethanol precipitation (*Protocol 3*, steps 4–5).

9 An alternative favoured by some workers is to substitute NaCl gradients for the sucrose gradient:

   (a) Prepare a 20% (w/v) NaCl solution in 50 mM Tris-HCl, pH 8.0, 1 mM EDTA.

   (b) Freeze and thaw once at −70 °C (−20 °C is insufficient to form a correct gradient).

   (c) Apply the sample and centrifuge at 270 000 **g** for 3 h at 15 °C in a Beckman SW40 rotor.

   (d) Purification of selected fractions from the gradient is by dialysis against sterile water overnight at 4 °C.

The λEMBL series of vectors (Chapter 9, Section 4.2) are designed for even simpler use. The internal fragment of these vectors is flanked by polylinkers. By digestion with two restriction enzymes the arms retain termini which can base pair with the digested target DNA whereas the internal fragment has termini which cannot ligate to the arms. The small oligonucleotide which separates the arms and the linker is readily removed by isopropanol precipitation.

Protocol 7

Preparation and purification of λEMBL3 arms

Reagents

- λEMBL3 DNA
- *Bam*HI and *Eco*RI restriction enzymes plus appropriate buffers
- 3 M sodium acetate, pH 5.5 (see *Protocol 3*)
- TE, pH 7.4 (10 mM Tris-HCl, pH 7.4, 1 mM EDTA, pH 8.0)

Method

1 Carry out a total *Bam*HI digestion of 50–100 μg of λEMBL3 DNA (*Protocol 1*).

2 Phenol extract and ethanol precipitate the DNA (*Protocol 3*).

3 Resuspend the *Bam*HI-digested DNA and prepare an *Eco*RI reaction solution (*Protocol 1*).

4 Phenol extract the *Bam*HI–*Eco*RI digested DNA.

5 To selectively remove the polylinker, precipitate the double-digested DNA with a 0.1 volume of 3 M sodium acetate pH 5.5 and 1 volume of isopropanol.

6 Incubate for 15 min at −20°C and centrifuge at 8 000 **g** in the microfuge for 10 min to pellet the DNA.

7 Resuspend the DNA at a concentration of 0.2–0.5 μg/μl in sterile TE pH 7.4 or water.

8 Store at 4°C or at −20°C for prolonged storage.

3.2 Cosmids

Cosmid libraries (Chapter 9, Section 3.6) can be seen as an extension of libraries prepared using λ as the vector. By inclusion of a small segment of λ into a plasmid, the *in vitro* packaging system can be used, giving high-efficiency transformation of *E. coli*. The size requirements for packaging remain unaltered and, as the plasmid element can be less than 4 kb long, the size of DNA that can be transferred may be over 40 kb long. Once transferred to *E. coli*, cosmids (lacking as they do, all λ functions) act as large plasmids. Because they typically carry antibiotic resistance genes, colonies which contain them grow on selective agar plates. They have the advantage, relative to phages, of containing a large segment of contiguous genomic DNA in each selected clone but also have the disadvantages of greater difficulty of storage and fewer units per plate, which makes their screening moderately more cumbersome.

When cosmid vectors are used, the methods involved are essentially the same as for plasmid vectors (see Section 3.3). Frequently, the cosmids are digested with two restriction enzymes which generate non-homologous termini (see *Protocol 2*). If these vectors ligate to form a dimer, it is not large enough to be packaged by the λ proteins and is therefore unable to be transferred to *E. coli*. For this reason the target DNA is frequently phosphatase-treated when cosmid libraries are generated (see *Protocol 8*). This ensures that only DNA which is contiguous in the genome is inserted into the vector. Given that cosmid vectors can incorporate fragments up to 40 000 bp and, as noted above, the visual estimate on an agarose gel of the number of molecules of small fragments can give a misleadingly low impression, particular care must be taken to eliminate DNA fragments that are smaller than those desired. Salt gradients are preferred by some when separating large DNA fragments (see *Protocol 6*, step 9) but any method in conjunction with phosphatase treatment of the DNA can be used effectively to diminish the number of cosmid clones with inappropriate inserts.

Protocol 8

Phosphatase treatment of DNA

Reagents

- BAP
- 10× BAP buffer (500 mM Tris-HCl, pH 8.0, 10 mM ZnCl$_2$)
- Proteinase K (100 μg/μl)
- TE, pH 8.0 (see *Protocol 3*)

- CIP
- 10× CIP buffer (200 mM Tris-HCl, pH 8.0, 10 mM MgCl$_2$, 10 mM ZnCl$_2$)

A. Phosphatase treatment with BAP

1   Prepare a reaction mixture containing appropriate amounts of 10× BAP buffer and DNA in a 100 μl volume.

2   Add 1 unit of BAP and incubate at 60 °C for 30 min. This high temperature suppresses any residual exonucleases in the enzyme preparation and is recommended when using BAP.

3   Stop the reaction by adding SDS to a final concentration of 0.1% (w/v) and proteinase K to 100 μg/μl. Incubate at 37 °C for 30 min.

4   Purify the phosphatase-treated DNA by phenol extraction and ethanol precipitation (*Protocol 3*) and resuspend the DNA in sterile TE, pH 8.0, or water.

5   Because of the stability of the enzyme and the consequences of its persistence in an experiment further purifications of the DNA are often performed before ethanol precipitation. These include extra phenol extractions or silica binding (Chapter 6, Section 2.1).

B. Phosphatase treatment with CIP

1   Set up a reaction mixture as in step A1 with 10× CIP buffer.

2   Add 1 unit of CIP and incubate at 37 °C for 30 min.

3   Stop the reaction by heating the mixture at 75 °C for 10 min (CIP is heat-labile whereas BAP is not) before purifying the phosphatase-treated DNA by phenol extraction and ethanol precipitation (*Protocol 3*).

Note

Shrimp alkaline phosphatase is also available from commercial suppliers and, being even more heat-labile than CIP, is often used when absolutely no carry-over of inactivated phosphatase into later manipulations can be tolerated. Use with a 10× buffer comprising 20 mM Tris-HCl, pH 8.0, 100 mM MgCl$_2$ and stop the reaction by heating at 65°C for 15 min.

3.3 Plasmids

There is a very wide choice of plasmids that can be used for genetic engineering experiments (see Chapter 9, Section 3). When used as vectors they are digested at one locus either by a single restriction enzyme or by two at a multi-cloning site to achieve insertion of target DNA in a defined orientation. The digestions are carried out as in *Protocols 1* and *2*. When digestions are verified as complete and correct by agarose gel electrophoresis, the DNA is phenol extracted and ethanol precipitated (*Protocol 3*). For many experiments it is advisable to treat the digested vector DNA with phosphatase (*Protocol 8*). By removing the 5'-phosphates from the plasmids, they can no longer be circularized by DNA ligase (see Section 4.1). This reduces the background of colonies which do not contain recombinant molecules in an experiment. It should be noted, however, that undigested supercoiled plasmids can transform *E. coli* with an efficiency approximately 50-fold greater than linearized DNA. As remnants of undigested plasmid DNA are frequently undetected by routine agarose gel electrophoresis, their contribution to background problems in cloning experiments should not be underestimated. If they persist after alkaline phosphatase treatment, more extensive digestion by the restriction enzyme or, ultimately, purification of the linear form of the plasmid by centrifugation through a sucrose gradient is recommended.

After alkaline phosphatase treatment it is imperative to remove all traces of the enzyme as it will impede further steps in the generation of recombinant molecules.

4 Construction of recombinant molecules

The previous sections have described how to prepare appropriately digested target and vector DNA. The final step of linking these together in a recombinant molecule prior to transfer to *E. coli* can be achieved in different ways that are outlined below.

4.1 Ligation

The purification of DNA ligase enzymes (5, 6) was important in developing the original concept of recombinant DNA. This enzyme links fragments of DNA to each other in a covalent manner (*Protocol 9*). The essential requirement for this ligation is that the DNA fragments present a 5'-phosphate group in close proximity to a 3'-hydroxyl group. When protruding complementary termini are present on the target and vector molecules they provide an obvious docking mechanism to bring the 5'- and 3'-ends of molecules together. However, ligation can also occur when there is no obvious mechanism for prior adhesion of fragments. Blunt-end ligation is not as efficient as ligation of complementary cohesive termini but it occurs readily when a higher concentration of DNA ligase is provided. It is often the method of choice for joining DNA molecules containing non-complementary protruding termini after these termini have been modified (*Protocol 10*). *Figure 1* shows the results of both cohesive-end and blunt-end ligation reactions. Ligation of DNA fragments with complementary ends (in this example, fragments of λ DNA cut with *Hind*III) is seen to occur when catalysed by a minimum of 0.01 units of T4 DNA ligase (*Figure 1A*, lane 2). When compared with the unligated DNA (*Figure 1A*, lane M), larger molecular weight DNA molecules can be seen. Greater ligation efficiency occurs when the amount of ligase is increased ranging from 0.025 units to 0.1 unit (*Figure 1A*, lanes 3–6). In con-

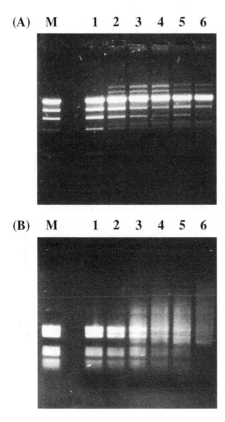

Figure 1 DNA ligation analysis. (A) Cohesive-end ligation. Lane M, 100 ng of λ DNA digested with *Hind*III (λ*Hind*III); lane 1, λ*Hind*III + 0.005 units (U) T4 DNA ligase; lane 2, λ*Hind*III + 0.01 U T4 DNA ligase; lane 3, λ*Hind*III + 0.025 U T4 DNA ligase; lane 4, λ*Hind*III + 0.05 U T4 DNA ligase; lane 5, λ*Hind*III + 0.075 U T4 DNA ligase; and lane 6, λ*Hind*III + 0.1 U T4 DNA ligase. (B) Blunt-end ligation. Lane M, 100 ng of pBR322 DNA digested with *Hae*III (pBR322*Hae*III); lane 1, pBR322*Hae*III + 0.05 U T4 DNA ligase; lane 2, pBR322*Hae*III + 0.1 U T4 DNA ligase; lane 3, pBR322*Hae*III + 0.25 U T4 DNA ligase; lane 4, pBR322*Hae*III + 0.5 U T4 DNA ligase; lane 5, pBR322*Hae*III + 0.75 U T4 DNA ligase; and lane 6, pBR322*Hae*III + 1.0 U T4 DNA ligase.

trast, the lower efficiency of blunt-end ligations is shown in *Figure 1B*. Ligation of pBR322 fragments with blunt ends generated by *Hae*III digestion occurs only when catalysed by a minimum of 0.25 units of T4 DNA ligase (*Figure 1B*, lane 3).

Protocol 9

DNA ligation

Reagents

- 10 × ligation buffer (0.66 M Tris-HCl, pH 7.6, 50 mM MgCl$_2$, 50 mM DTT, 10 mM ATP)
- T4 DNA ligase

Method

1 The volume of the ligation mixture and the DNA concentration depend on the type of ligation experiment. Use a 10 μl reaction with DNA at a concentration of > 100 ng/μl for concatamer ligation products, or a 10 μl reaction with DNA at < 10 ng/μl for circular ligation products.

2 Add T4 DNA ligase. For a cohesive-end ligation add 0.25 units enzyme/μg of DNA, and for a blunt-end ligation add 2.5 units/μg.

3 Incubate the reaction mixture at 15 °C for 1–16 h. Simple cohesive-end ligations are usually complete in 1–2 h.

4 Analyse for correct and complete ligation by gel electrophoresis with unligated material as a marker, or by transformation if the resulting DNA contains vector sequences.

Protocol 10

Preparation of blunt-ended DNA fragments

Reagents

- Klenow DNA polymerase
- 10× Klenow reaction buffer (500 mM Tris-HCl, pH 7.5, 100 mM MgCl$_2$, 10 mM DTT)
- Stock solutions (e.g. 20 mM) of each dNTP
- T4 DNA polymerase
- 10× T4 pol reaction buffer (500 mM Tris-HCl, pH 8.8, 50 mM MgCl$_2$, 50 mM DTT)

A. Filling in 5′-overhangs with Klenow polymerase

1 Prepare a reaction mixture containing DNA, 1× Klenow reaction buffer and 20 μM each dNTP.

2 Add 1 unit Klenow DNA polymerase and incubate at 15–37 °C for 15–30 min.

3 Stop the reaction by heating at 75 °C for 10 min.

4 The progress of the reaction can be monitored by a simple ligation experiment, followed by gel electrophoresis. DNA that originally had complementary termini should not now be ligated by low amounts of T4 ligase (*Protocol 9*); ten times the amount of ligase should now be required to achieve the necessary blunt-end ligation.

B. Removal of 3′-overhangs by the exonuclease reaction of T4 DNA polymerase

1 Prepare a reaction mixture containing DNA, 1 × T4 DNA pol reaction buffer and 20 μM each dNTP.

2 Add 1 unit of T4 DNA polymerase and incubate at 15 °C for 15–30 min.

3 Stop the reaction by heating at 75 °C for 10 min.

4 Analyse the reaction by ligation of a sample.

The standard ligation reaction outlined in *Protocol 9* is clearly a simple one. However, experience shows that it frequently does not work efficiently. Some simple precautions are suggested. Principal among these is the usefulness of showing that the individual components can ligate to themselves. This control allows the researchers to establish which fragment is the source of the problem and show, by use of known materials, that the ligation enzyme and buffer are capable of carrying out the ligation.

If the problem is with one of the DNA fragments then either the restriction enzyme used in the digestion contained some exonuclease whose activity has made cohesive ligation inefficient, or the fragment is in a solution that contains either organic solvents or other contaminants such as particles from agarose gel. A mixing experiment using DNA fragments that are known to be capable of ligation will distinguish between these two possibilities.

The result of a ligation depends on the relative concentrations of the DNA molecules in the solution. Two types of product from ligation can be envisaged: linear molecules, which are multimers (concatamers) of the component fragments, or circular molecules. When cloning with λ or cosmid vectors, the former are required. When using plasmid vectors the latter are preferred. If the concentration of DNA termini in a solution is very high then concatamers are more likely to be formed because of the statistical probability of two molecules being joined rather than one molecule circularizing and linking to itself. If solutions of DNA fragments are dilute, the converse is true. The length of the DNA fragment also influences the outcome of the ligation step. Mathematical treatments of these factors have been presented (7). The practical consequences of these considerations are that the total DNA concentration in a plasmid cloning experiment should be maintained at less than 1 µg/ml and in a λ cloning experiment at over 100 µg/ml.

4.2 Linkers and adapters

Although blunt-end ligation can work effectively, it is much simpler to ligate large DNA molecules which have cohesive termini. Such termini may be added artificially to DNA fragments by the addition of small oligonucleotides, the sequence of which includes a restriction site. These oligonucleotides are of two types: linkers and adapters.

Linkers are complementary oligomers which form small double-stranded DNA fragments which include a restriction enzyme site. They are ligated to blunt-end DNA by DNA ligase. Because of the high concentration of these small molecules present in the reaction, the ligation is very efficient when compared with blunt-end ligation of large molecules. The terminal restriction enzyme site is generated by digestion with the appropriate enzyme. Obviously, sites for the enzyme may be present in the target DNA fragment and, therefore, any such sites are protected by prior methylation. This requirement limits the use of linkers to those for which modification enzymes (methylases) are available which correspond to the restriction enzyme target site (4).

Adapters are similar to linkers in that they are small oligonucleotides which are blunt-end ligated to the target DNA. They are not perfectly double-stranded, however, and are synthesized to present cohesive termini without any digestion. In this way, the target DNA is ready for ligation to the complementary restriction enzyme termini of the vector.

Protocol 11

Addition of linkers to blunt-end DNA molecules

Equipment and reagents

- T4 DNA ligase and 10× ligation buffer (see *Protocol 9*)
- Bio-Gel A–50 m column (Bio-Rad)
- Restriction enzyme plus appropriate buffer

Method

1 Prepare a ligation mixture containing DNA and linkers in a 50 μl volume (*Protocol 9*). Keep a 1:1 ratio of linkers to DNA (e.g. 1 μg DNA to 1 μg linkers) because this ensures a large molar excess of linkers.

2 Add 5 units T4 DNA ligase and incubate for 16 h at 15 °C.

3 Stop the reaction by heating at 70 °C for 10 min.

4 Increase the volume of the ligation mixture to 100 μl by adding 10 μl of the appropriate 10× restriction enzyme buffer and 40 μl water.

5 Add 100 units of restriction enzyme and incubate at 37 °C for 12 h.

6 Heat at 70 °C for 10 min to inactivate the restriction enzyme.

7 Separate the large DNA with linkers at its termini from digested linkers by size fractionation on a Bio-Gel A–50 m column (*Protocol 12*).

8 Purify and concentrate the large DNA by ethanol precipitation (*Protocol 3*, steps 4 and 5).

Notes

(a) If the DNA contains internal restriction sites for the enzyme being used to cleave the ligated linkers, then these sites must first be blocked by treatment with the appropriate DNA methylase. Set up a 20 μl reaction containing the DNA, methylase, the appropriate buffer (obtained from the supplier of the methylase) and 80 μM *S*-adenosylmethionine. Incubate at 37 °C (for most methylases) for 1 h then recover the DNA by phenol extraction and ethanol precipitation (*Protocol 3*).

(b) Adapters are becoming more favoured as a method of adding complementary termini to blunt-end DNA molecules. The procedure is similar to that of linkers. However, no restriction enzyme digestion is required, thereby also removing the need to methylate the target DNA. The adapters containing single-strand complementary ends are ligated on to the DNA (step 1), and phosphorylated with T4 polynucleotide kinase directly in the ligation mixture. Excess adapters can be removed by gel filtration (*Protocol 12*).

Protocol 12

Removal of excess linkers or adapters

Equipment and reagents

- Disposable 5 ml or 10 ml plastic pipette
- Sterile glass wool
- Bio-Gel A–50 m (Bio-Rad)
- Column buffer (10 mM Tris-HCl, pH 7.5, 100 mM NaCl, 1 mM EDTA)

Method

1   Build the column in a sterile 5 ml or 10 ml plastic pipette. Plug the bottom of the pipette with sterile glass wool.

2   Pre-swell the Bio-Gel A–50 m in column buffer for 1 h at room temperature. Pour the resin into the pipette giving a bed volume of 5 ml. Wash the column with 50 ml column buffer. This removes a ligase inhibitor from the resin that may interfere with further cloning reactions.

3   Load the sample on to the column, wash with more column buffer and collect between 20 and 30 100-µl fractions.

4   Thorough washing of the column with more buffer allows reuse.

5   Analysis of the fractions can be carried out by various methods. Simple electrophoresis of a sample from each fraction may show which fractions contain the modified DNA. Alternatively trace amounts of radiolabel can be incorporated into the linkers (see Volume II, Chapter 4) and the fractions analysed using a scintillation counter. This shows two peaks of radioactivity: an earlier peak containing the modified DNA and a final large peak containing excess labelled linkers/adapters. The column may also be equilibrated before use with known DNA size markers. Analysis by gel electrophoresis will then determine which fraction contains DNA of a particular size.

4.3 Tailing

The addition of long homopolymers of dNTPs to the ends of molecules was particularly popular in the initial years of cDNA cloning prior to the ready availability of linkers and adapters. TdT is an unusual polymerizing enzyme in that it does not require a template: if a dNTP is provided then TdT will add it to any available 3′-hydroxyl site which serves as a primer. The kinetics of addition vary for each dNTP. Because of inhibiting secondary structures that result from polymers of deoxyguanosine, elongation with this nucleotide is self-limiting to approximately 30 additions. The optimum primer is a protruding 3′-sequence but blunt-end fragments can be effective in the presence of the cacodylate-cobalt buffer which allows transient single strands to occur at the ends. A popular combination is the use of d(G) tails on the vector and highly efficient d(C)-tailing of the target DNA.

Protocol 13

Homopolymer tailing and annealing reactions

Reagents

- 10× tailing buffer (1 M sodium cacodylate, pH 7.0, 10 mM CoCl$_2$, 1 mM DTT)[a]
- Stock solution (e.g. 20 mM) of the appropriate dNTP
- TdT
- 10× annealing buffer (500 mM Tris-HCl, pH 8.0, 10 mM EDTA, 1 M NaCl)

Method

1. Prepare a tailing mixture of 50 μl containing DNA, 1× tailing buffer and a 20 μM concentration of the nucleotide to be added to the DNA.

2. Add 10 units of TdT and incubate at 37 °C for 10 min. Under these conditions the enzyme will add 20 nucleotides to the 3′-end of the DNA in 10 min if the reaction includes 4 pmol of DNA. Trial experiments can be carried out with trace amounts of radioactive dNTPs to determine the extent of addition of nucleotides on to the 3′-ends of the DNA.

3. Stop the reaction by heating at 75 °C for 10 min. Purify the DNA by phenol extraction and ethanol precipitation (*Protocol 3*). It is also useful to dialyse the DNA against sterile water for 1–2 h to remove components from the tailing mixture that may inhibit bacterial transformation (*Protocol 3*, step 6).

4. Prepare an annealing mixture containing vector and target DNAs with complementary homopolymer tails in 100 μl of 1× annealing buffer. The DNA concentration and ratio of vector to target depends on the nature of the cloning experiment (*Protocol 9*). Heat the sample at 70 °C for 30 min in a water bath. Turn off the heat and allow the sample to cool in the water bath to room temperature over a period of 7–12 h.

5. The sample is now ready for transformation.

[a] Extreme care is required when making up the 10× tailing buffer. Add the chemicals in the order stated to prevent precipitation.

4.4 T/A cloning of PCR products

With the advent of PCR, one of the most common cloning reactions in the current molecular biology laboratory is the ligation of PCR-amplified DNA fragments to plasmid vectors and subsequent introduction into *E. coli* cells. Although several methods exist for subcloning PCR products (e.g. blunt-end ligation or the incorporation of restriction enzyme recognition sites into the PCR primers: see Volume II, Chapter 7), the T/A procedure is the most popular choice. This method relies on the TdT activity of some thermophilic polymerases including *Taq* polymerase. This activity means that in a typical amplification reaction the majority of the PCR products have an A residue as a 3'-overhang. These DNA fragments can then be ligated to plasmid vectors which have a complementary 3'-T overhang. The plasmid vector can be prepared by digestion with a restriction enzyme which produces a 3'-T overhang (e.g. *BspCI*) or by incubating blunt-end plasmid DNA molecules with *Taq* polymerase in the presence of excess dTTP (*Protocol 14*). Linear vector DNA with 3'-T overhangs are also available commercially (see Chapter 9, Section 3.7).

Protocol 14

T/A cloning of PCR products

Reagents

- *Taq* DNA polymerase
- 10× *Taq* DNA polymerase buffer (200 mM Tris-HCl, pH 8.5, 15 mM MgCl$_2$, 500 mM KCl)
- Stock solution (e.g. 20 mM) of dTTP
- T4 DNA ligase and 10× ligation buffer (see *Protocol 9*)

Method

1 Prepare a 50 μl reaction mixture containing 1.5 μg blunt-end plasmid DNA, 1× *Taq* DNA polymerase buffer and 20 μM dTTP.

2 Add 1 unit of *Taq* DNA polymerase and incubate at 70 °C for 1 h.

3 Purify the plasmid DNA by phenol extraction and ethanol precipitation (*Protocol 3*) and resuspend at a concentration of 50 ng/μl in sterile water.

4 Prepare a ligation mixture containing 0.25 units of T4 DNA ligase and an equimolar ratio of the plasmid DNA and PCR products (see *Protocol 9*). Incubate at 15 °C for 1–16 h.

5 Conclusion

The generation of recombinant molecules requires a combination of methods that are deceptively complex. The quality of the materials used at all stages, whether purchased or generated, is of the utmost importance. Although it is true that 'only one correct clone is needed', inefficiency at any of the steps described is usually severely punished. Attention to detail, which is the essence of good science, is well rewarded with libraries capable of providing answers to almost all questions in molecular biology.

Acknowledgements

The comments and recommendations of our colleagues and research group members are gratefully acknowledged.

References

1. Watson, J. D. and Crick, F. H. C. (1953). *Nature* **171**, 946.
2. Crick, F. H. C. and Watson, J. D. (1954). *Proc. R. Soc. Lond. Ser. B* **223**, 80.
3. http://www.neb.com/rebase/rebase.html
4. Brown, T. A. (ed.) (1998). *Molecular biology labfax*, 2nd edn, Vol. 1. Academic Press, London.
5. Weiss, B., Jacquemin-Sablon, A., Live, T. R., Fareed, G. C., and Richardson, C. C. (1968). *J. Biol. Chem.* **243**, 4543.
6. Panasenko, S. M., Cameron, J. R., Davis, R. W., and Lehman, I. R. (1977). *J. Biol. Chem.* **253**, 4590.
7. Dugaiczyk, A., Boyer, H. W., and Goodman, H. M. (1975). *J. Mol. Biol.* **96**, 174.

Chapter 8

Generation and identification of recombinant clones

T. A. Brown

Department of Biomolecular Sciences, University of Manchester Institute of
Science and Technology, Manchester M60 IQD, UK

1 Introduction

The central steps in a gene cloning experiment are the introduction of recombinant DNA molecules synthesized *in vitro* into host *Escherichia coli* cells and the subsequent identification of recombinant clones. In outline the procedure is as follows:

(1) Bacteria and DNA are mixed together and some of the cells take up DNA molecules.

(2) The cells are plated on to a selective medium (or a series of media) that enables recombinant colonies (i.e. those that contain recombinant DNA molecules) to be identified. There are several different selection strategies; the appropriate one depends on the type of vector that is being used (see also Chapter 9).

This chapter begins with methods for introducing recombinant DNA molecules into bacteria and then provides details for recombinant selection with the most important vectors. The protocols deal solely with the use of *Escherichia coli* as the host organism, *E. coli* being the host that is used for all basic gene cloning experiments.

2 Introduction of DNA into *E. coli* cells

There are three different approaches to the introduction of recombinant DNA molecules into *E. coli*:

- Uptake of plasmid or bacteriophage molecules by chemically treated cells
- Uptake of plasmid or bacteriophage molecules by electroporation, a physical treatment that uses an electric charge to generate transient breaks in the cell exterior, through which the DNA enters
- *In vitro* packaging, which involves construction of infective phage particles that contain the recombinant DNA molecules

The natural process by which bacteria take up naked DNA molecules is called transformation. This term is also used to describe the uptake of recombinant DNA molecules by chemically treated cells and by electroporation, and is sometimes even extended to include *in vitro* packaging. Transfection is sometimes substituted for transformation if the process involves phage rather than plasmid DNA. None of these usages are strictly correct but they are now embedded in molecular biology jargon and few gene cloners worry about the precise meaning of the terms.

2.1 Uptake of DNA by chemically treated cells

In nature, transformation is probably not a major process by which bacteria obtain genetic information (1). This is reflected by the fact that in the laboratory only a few bacteria (notably *Bacillus* and *Streptococcus* species) can be transformed with ease. Most species, including *E. coli*, take up only limited amounts of DNA under normal circumstances and have to undergo a chemical pretreatment before they can be transformed efficiently. Cells that have undergone such a treatment are said to be competent.

2.1.1 Preparation of competent cells

Methods for preparing competent *E. coli* cells derive from the work of Mandel and Higa (2) who developed a simple treatment based on soaking the cells in cold $CaCl_2$. Why this treatment is effective is not known. Their original experiments were with λ DNA but the procedure was quickly shown to be applicable to plasmid (3) and chromosomal (4) DNA. *Protocol 1* is a simple and easy method based on this original procedure.

Protocol 1

Preparation of competent *E. coli* cells

Reagents

- Appropriate strain of *E. coli*
- LB medium (10 g bacto-tryptone, 5 g bacto-yeast extract, 10 g NaCl per 1 l). Check the pH and adjust to 7.0–7.2 with NaOH.

Sterilize by autoclaving at 121 °C, 103.5 kPa (15 lb/in^2), for 20 min
- Ice-cold, sterile 50 mM $CaCl_2$

Method

1 Inoculate 5 ml of LB medium with a colony of *E. coli*. Incubate overnight at 37 °C with shaking.

2 Remove 300 μl of the overnight culture and inoculate 30 ml of fresh LB medium. Incubate at 37°C with shaking until the OD at 550 nm reaches 0.4–0.5. This should take about 2 h.

Protocol 1 continued

3   Centrifuge the cells at 6000 g for 10 min at 4 °C in a pre-cooled rotor. It is important that from this point onwards the cells are not allowed to warm up.

4   Discard the medium and resuspend the cells in 15 ml ice-cold, sterile 50 mM $CaCl_2$. Leave on ice for 15 min with occasional shaking.

5   Centrifuge the cells as in step 3 and discard the $CaCl_2$.

6   Resuspend in 3 ml ice-cold, sterile 50 mM $CaCl_2$. The cells are now competent.

Notes

(a) The growth stage that the cells have reached is critical. At less than OD 0.4–0.5 there will be insufficient cells for transformation. At higher cell densities the procedure will not result in a good yield of competent cells.

(b) The procedure can be scaled up to as much as 1 l of starting cells. The initial pellet should be resuspended in a one-half volume of $CaCl_2$ and the competent cells taken up in a 0.1 volume of $CaCl_2$.

This procedure is very reliable and works well with most strains of *E. coli*. The cells can be used immediately or stored at −80°C after the addition of 15% (v/v) glycerol. Stored cells retain their competence for several months, provided that they are not allowed to warm up. They should be stored in 300 μl aliquots so that stocks do not have to be repeatedly thawed.

Cells prepared by the method shown in *Protocol 1* should yield more than 10^7 transformants/μg of supercoiled plasmid DNA, which is satisfactory for most standard cloning procedures. However, for some applications the amount of available DNA is so low that higher transformation efficiencies are required. A combination of highly-transformable strain (e.g. DH1, see *Appendix 1*) and more sophisticated technique can yield competent cells that provide a 50-fold improvement in transformation efficiency (5).

Protocol 2

Preparation of highly-transformable *E. coli*

Reagents

In this procedure it is important to use pure, quality reagents and the cleanest glassware

- Highly transformable strain of *E. coli*, (e.g. DH1, see *Appendix 1*)

- SOB agar (20 g bacto-tryptone, 5 g bacto-yeast extract, 0.5 g NaCl, 15 g bacto-agar per 1 l). Check the pH and adjust to 7.0–7.2 with NaOH. Sterilize by autoclaving at 121°C, 103.5 kPa (15 lb/in²), for 20 min

- SOB medium + 20 mM $MgSO_4$. Prepare SOB medium (2 g bacto-tryptone, 0.5 g bacto-yeast extract, 0.05 g NaCl per 90 ml), check the pH, adjust to 7.0–7.2 with NaOH, and sterilize by autoclaving at 121 °C, 103.5 kPa (15 lb/in²), for 20 min. Immediately before use, add $MgSO_4$ to a final concentration of 20 mM from an autoclaved 1 M stock solution

Protocol 2 continued

- Transformation buffer A (contains 10 mM MES, pH 6.3, 45 mM $MnCl_2$, 10 mM $CaCl_2$, 100 mM KCl, 3 mM hexaminecobalt chloride) or transformation buffer B (10 mM potassium acetate, pH 7.5, 45 mM $MnCl_2$, 10 mM $CaCl_2$, 100 mM KCl, 3 mM

  hexaminecobalt chloride, 10% [v/v] glycerol)

- DD medium (1.53 g DTT, 9 ml DMSO, 100 μl 1 M potassium acetate, pH 7.5, 900 μl water) or DMSO

Method

1 Make a streak plate (Chapter 2, *Protocol 2*) on SOB agar with the *E. coli* strain. Ideally, the inoculum should come from a frozen stock.

2 Incubate the plate overnight at 37 °C and then transfer five small colonies (about 2 mm diameter) into 1 ml SOB + 20 mM $MgSO_4$.

3 Gently vortex to break up the colonies, then inoculate into 100 ml SOB + 20 mM $MgSO_4$ in a 1-l flask.

4 Incubate at 37 °C with shaking until the culture reaches 10^8 cells/ml. This will correspond to about 2 h incubation and a final OD at 550 nm of 0.4–0.5.

5 Split the culture into 50-ml portions and transfer to pre-cooled centrifuge tubes. Cool on ice for 10 min, then centrifuge at 6000 **g** for 10 min at 4 °C.

6 Discard the medium, including the last traces which can be removed by careful aspiration.

7 Gently resuspend the cells from each tube in 20 ml of ice-cold, sterile transformation buffer. The composition depends on whether the cells are to be used immediately or stored:

   (a) If the cells are to be used immediately then use transformation buffer A

   (b) If the cells are to be stored at −80 °C then use transformation buffer B.

8 Stand the cells on ice for 10 min, then centrifuge as in step 5.

9 Remove the buffer, including the final traces, and carefully resuspend each pellet in 4 ml of the appropriate transformation buffer.

10 If the cells are to be used be immediately:

   (a) Add 140 μl of filter-sterilized DD to each suspension and stand on ice for 15 min.

   (b) Add a further 140 μl of DD and stand on ice for a further 15 min. The cells are now ready to use.

11 If the cells are to be stored:

   (a) Add 140 μl DMSO to each suspension and stand on ice for 15 min.

   (b) Add a further 140 μl DMSO, quickly dispense 50-μl aliquots into chilled microfuge tubes and freeze in liquid nitrogen. Store at −80 °C.

As an alternative to preparing your own competent cells they can be purchased from any one of several suppliers. The quality is generally good and a range of different *E. coli* strains is available, but competent cells are so easy to prepare in the laboratory that few molecular biologists purchase commercial cells on a regular basis. The exception is when the host strain is heavily mutated, for example to prevent rearrangement of cloned DNA (see *Appendix 1*), a side-effect of these mutations being that the bacteria are difficult to culture and are somewhat refractory to standard procedures for rendering them competent.

2.1.2 DNA uptake by competent cells

Competent cells are induced to take up DNA by a short heat shock. As with the $CaCl_2$ treatment, the biological basis to the technique is not understood. It is, however, relatively efficient.

Protocol 3

Uptake of DNA by competent cells

Equipment

- 42 °C water bath or incubator

Method

1 Add a suitable amount of DNA[a] to 300 µl of competent cells (from *Protocol 1*) or 50 µl of competent cells (from *Protocol 2*).

2 Leave on ice for 30 min.

3 Transfer to 42 °C for 2 min.

4 Return to ice. Move fairly rapidly to the next stage of the procedure (*Protocol 4* or Section 3).

[a] The ideal amount of DNA to add depends on various factors, the most important of which is the transformation frequency of the vector. Some vectors have high transformation frequencies and so give many transformants per nanogram of DNA, whereas others have a lower transformation frequency and give relatively few transformants. The number of transformants does not usually increase at DNA amounts above 2 ng for the volumes of competent cells used in this procedure. It is therefore better to use several aliquots of competent cells, and subsequently a larger number of agar plates for recombinant selection, than to add excessive amounts of DNA into one tube. The volume of DNA should be less than 5% that of the competent cells (e.g. no more than 15 µl of DNA per 300 µl of competent cells), partly because the $CaCl_2$ concentration must be maintained at 50 µM or above (extra could be added to compensate) and partly because components of the ligation mixture may interfere with DNA uptake.

2.1.3 Expression of antibiotic resistance genes carried by plasmid vectors

Several plasmid cloning vectors carry genes for antibiotic resistance and are plated on to antibiotic media for transformant and recombinant selection (Section 3.1). If cells containing these vectors are plated out immediately after DNA uptake, the yield of colonies will be low as there will not have been time for expression of the antibiotic resistance genes before the cells actually encounter the inhibitor. To circumvent this problem the cells are usually incubated in broth for a short period before being plated out.

Protocol 4

Plasmid expression

Reagents

- LB medium (see *Protocol 1*)

Method

1   Add 0.5 ml LB medium, pre-warmed to 37 °C, to each tube of cells immediately after the heat shock (*Protocol 3*, step 3).

2   Incubate at 37 °C for 1 h.

3   Return to ice.

2.1.4 Troubleshooting

The methodology for uptake of DNA by chemically treated cells is robust and problems are rarely experienced if the protocols are carefully followed. Unfortunately, if an error does occur the result is not apparent until the end of the experiment, when the cells are plated and no or very few transformants are obtained. It is therefore worth bearing in mind the commonest reasons for failure in order to avoid these pitfalls when the experiment is carried out.

Total failure, leading to a complete absence of transformants, is uncommon. The usual explanations are:

- *A trivial error occurred* and the DNA was lost at an earlier part of the experiment.

- *The host strain is contaminated*. If this is suspected then follow the guide-lines in Chapter 2, Section 6.2 to track down and eliminate the cause(s).

A more common problem is low transformation efficiency, for which there are several explanations:

- *The cells were grown to the wrong stage*. The most important aspect of preparation of competent cells is to make sure the starting culture is neither too young nor too old. Do not be impressed by molecular biologists who judge the OD by eye: measure it properly!

- *The cells were allowed to warm up*. Cells that return to room temperature rapidly lose their competence. It is possible to prepare competent cells with a non-refrigerated bench-top centrifuge but transformation efficiencies will be low. Once out of the centrifuge the cells should be returned to ice and solutions added to the cells should be pre-cooled on ice.

- *The DNA was added in too high a volume.* Keep to the ratio of DNA to competent cells described above. If the DNA has to be in a large volume then adjust to 50 mM $CaCl_2$ by adding the appropriate volume from a 1 M stock.

- *The heat-shock was ineffective.* Two minutes at 42 °C is sufficient for cells in standard 1.5-ml microfuge tubes. If you are using different tubes then the time will have to be altered. In particular, a 50-ml centrifuge tube will require a longer period at 42 °C to provide the cells with the same heat exposure.

Identifying the reason for low transformation efficiency can be difficult and it should be borne in mind that, in general terms, high efficiencies are not always needed. As long as you obtain a sufficient number of recombinants you need not worry about the efficiency of your transformation system. This is not an excuse for poor technique but it is not time-effective to spend weeks optimizing a procedure if it is already providing you with the desired result.

2.2 DNA uptake by electroporation

Transformation frequencies 100–1000 times greater than are possible with chemically treated cells can be obtained by electroporation. The procedure requires special apparatus but is worth considering if many recombinants are required, for example, when a genomic or cDNA library is being prepared or if the desired recombinant DNA molecule is only a minor component of the DNA sample that is being used in the transformation.

The electroporation apparatus applies an electric field of approximately 12.5 kV/cm^2 in 5–10 ms pulses to the *E. coli* cell suspension. It is thought that the electric pulses induce the formation of small holes in the bacterial cell membrane, through which the DNA enters before the holes are repaired (6). Transformation efficiencies of up to 10^{10}/μg of supercoiled plasmid DNA have been reported, but this is with rigorously purified DNA and, in practice, it is rarely possible to obtain more than 10^9 transformants/μg DNA. It is also necessary to wash the cells repeatedly in water prior to electroporation because traces of growth medium can reduce the transformation efficiency.

Protocol 5

Uptake of DNA by electroporation

Equipment and reagents

- Electroporation apparatus plus cuvettes
- Appropriate strain of *E. coli*
- LB medium (see *Protocol 1*)
- Recovery medium. Prepare the basic recovery medium (2 g bacto-tryptone, 0.5 g bacto-yeast extract, 0.05 g NaCl per 100 ml), check the pH, adjust to 7.0–7.2

with NaOH and sterilize by autoclaving at 121 °C, 103.5 kPa (15 lb/in^2) for 20 min. Immediately before use, add aliquots of autoclaved or filter-sterilized stock solutions to give the following final concentrations: 2.5 mM KCl, 20 mM $MgSO_4$, 20 mM glucose

Protocol 5 continued

Method

1 Inoculate 5 ml of LB medium with a colony of *E. coli*. Incubate overnight at 37°C with shaking.

2 Add the entire overnight culture to 500 ml of fresh LB medium. Incubate at 37°C with shaking until the OD at 550 nm reaches 0.5–0.6. This should take about 2.5 h.

3 Centrifuge the cells at 6000 g for 10 min at 4°C in a pre-cooled rotor. It is important that from this point onwards the cells are not allowed to warm up.

4 Discard the medium and resuspend the cells in 500 ml ice-cold, sterile water.

5 Centrifuge the cells as in step 3, discard the water and resuspend the cells in another 500 ml ice-cold, sterile water.

6 Centrifuge the cells as in step 3, pour off the water and resuspend in a minimal volume of ice-cold, sterile water.[a]

7 Remove 100 µl of cells and add to the DNA sample.[b]

8 Place the mixture in a chilled electroporation cuvette and insert into the apparatus. Follow the operating instructions to set the appropriate electrical parameters and to carry out the electroporation.

9 Immediately after electroporation, remove the cuvette from the apparatus and transfer the cells into 1 ml of recovery medium. Incubate at 37°C with shaking for 1 h.

10 Continue with the next stage of the procedure (*Protocol 4* or Section 3).

[a] The ideal cell density is 10^{11}/ml, which can be achieved by resuspending the cell pellet in sufficient water to bring the volume up to 1 ml. The cells should be used immediately.

[b] Efficient electroporation requires a minimum of 5 pg DNA. With 100 ml cells at 10^{11} cells/ml the practical upper limit is 200 ng DNA. The DNA should be in a minimal volume, ideally 1 µl, to avoid reducing the cell density.

2.3 *In vitro* packaging

It is possible to transform *E. coli* cells with recombinant λ DNA molecules, using either chemical treatment or electroporation, but this is a relatively inefficient way of introducing phage DNA into host cells. Instead, infective λ phage particles are constructed in the test tube by the process called *in vitro* packaging. This requires a number of different proteins coded by the λ genome, which can be prepared at high concentration from cells infected with defective λ phage strains. Two different systems are in use. The first makes use of a pair of defective strains (*E. coli* BHB2688 and BHB2690; see *Appendix 1*), each carrying a mutation in one of the components of the phage protein coat (7). Infected cells synthesize and accumulate all the other components, but cannot assemble mature phage particles. However, a mixture of lysates from the two strains will contain all the required proteins and can be used for *in vitro* packaging. The second system makes use of *E. coli* SMR10, which has defective *cos* sites (8). This strain synthesizes all λ proteins but cannot assemble phage particles *in vivo*, as it does not recognize its own DNA as a substrate for packaging. However, a lysate will package recombinant λ DNA molecules that carry suitable *cos* sites.

Procedures for preparing packaging extracts from *E. coli* BHB2688 and BHB2690 are given in *Protocol 6*. This two-strain method is generally more reliable than the one-strain system and is recommended if packaging extracts are to be prepared in the laboratory. However, the preparation of effective packaging extracts is difficult and time-consuming whichever system is used, and it is advisable to purchase ready-made extracts from a commercial supplier (e.g. Stratagene Gigapack III Gold packaging extract). In general, the efficiency of commercial extracts is lower than those made in the laboratory (assuming you have mastered the technique), but still satisfactory. The use of laboratory or commercial extracts in *in vitro* packaging is described in *Protocol 7*.

Protocol 6

Preparation of λ packaging extracts[a]

Equipment and reagents

- Water baths or incubators at 30 °C, 38 °C, 42 °C and 43 °C
- Ti50 or Ti75 rotor (Beckman)
- Sonicator with a microtip device
- *E. coli* BHB2688
- *E. coli* BHB2690
- LB agar plates (10 g bacto-tryptone, 5 g bacto-yeast extract, 10 g NaCl, 15 g bacto-agar per 1 l). Check the pH and adjust to 7.0–7.2 with NaOH. Sterilize by autoclaving at 121 °C, 103.5 kPa (15 lb/in^2) for 20 min
- LB medium (see *Protocol 1*)

- BHB2688 resuspension buffer (50 mM Tris-HCl, pH 7.6, 10% [w/v] sucrose)
- Lysozyme solution (2 mg/ml in BHB2688 resuspension solution)
- M1 buffer (add together in the following order: 0.55 ml water, 5 µl β-mercaptoethanol, 30 µl 0.5 M Tris-HCl, pH 7.6, 1.5 ml 0.05 M spermidine plus 0.1 M putrescine, pH 7.0, 45 µl 1 M MgCl$_2$, 375 µl 0.1 M ATP)
- BHB2690 resuspension buffer (20 mM Tris-HCl, pH 8.0, 3 mM MgCl$_2$, 10 mM β-mercaptoethanol, 1 mM EDTA)

A. Preparation of a freeze–thaw lysate of *E. coli* BHB2688

1   Remove a single colony of *E. coli* BHB2688 from a stock culture and touch on to the surface of an LB agar plate that has been prewarmed to 42 °C. Streak onto an LB agar plate at room temperature (see Chapter 2, *Protocol 2* for the streak plate method). Incubate the first plate at 42 °C and the second at 30 °C. The strain should not grow at 42 °C: if it does then check another colony and, if necessary, obtain a new stock.

2   Remove the cells from the 30 °C plate and resuspend in a minimal volume of LB medium pre-warmed to 30 °C. Transfer the entire resuspension, in equal aliquots,

to 6×400 ml of pre-warmed LB in 2-l flasks. Incubate at 30 °C with shaking until the absorbance at 600 nm reaches 0.3. Transfer the flasks to a 43 °C water bath or incubator and shake gently for 20 min to induce the λ prophage, then transfer to 38 °C and shake vigorously for 3 h.[b]

3 Chill the cultures in ice-water, then centrifuge for 10 min at 5000 **g** at 4 °C. Decant the supernatants and resuspend each pellet in 1.2 ml cold BHB2688 resuspension buffer.[c]

4 Add 120 μl lysozyme solution to each resuspended pellet, mix, then place the tube in liquid nitrogen. Either proceed immediately to step 5 or store the extract at −70 °C.

5 Thaw each sample slowly on ice (this should take about 1 h). Add 0.5 ml M1 buffer to each one, shake gently, and then centrifuge at 2 °C for 30 min at 120 000 **g** in a pre-cooled Ti50 or Ti75 rotor.

6 Pool the supernatants, then distribute 50-, 100-, and 250-μl aliquots into cold tubes, freeze in liquid nitrogen, and store at −70 °C.

B. Preparation of a sonic extract of *E. coli* BHB2690

1 Test the strain as described in Part A, step 1.

2 Grow a single 400-ml culture and induce the λ prophage as described in Part A, step 2. After induction, shake vigorously for 2 h at 38 °C.

3 Chill the culture in ice-water, then centrifuge for 10 min at 5000 **g** at 2 °C. Decant the supernatant and resuspend the pellet in 1.5 ml cold BHB2690 resuspension buffer.[c]

4 Transfer the suspension to a clear plastic tube. Sonicate while cooling within an ice-salt mixture, approximately 15 times for 5 s at full power using a microtip soni-cator. The solution has to be ice-cold at all times and there should be no foaming. Sonication is complete when the solution is translucent.

5 Centrifuge the solution for 10 min at 8000 **g** in a Sorvall SS-34 rotor. There should be a very small pellet: if not, the sonication has been insufficient. Add 0.6 ml M1 buffer to the supernatant, distribute aliquots of 20-, 50- and 100-μl into cold tubes, freeze in liquid nitrogen, and store at −70 °C.

[a] See also Volume II, Chapter 2, *Protocol 8*.

[b] These incubations induce the prophage version of the defective λ genome carried by the *E. coli* strain, resulting in synthesis of the λ proteins.

[c] The cell suspension must be kept cold at all times so carry out these manipulations on ice.

Protocol 7

In vitro packaging[a]

Equipment and reagents

- Packaging extracts (*Protocol 6*) or commercial preparations (e.g. Stratagene Gigapack III Gold packaging extract)

- SM buffer (50 mM Tris-HCl, pH 7.5, 100 mM NaCl, 8 mM MgSO$_4$, 0.01% gelatin)
- 22 °C water bath or incubator

Method

1 Mix together 1–2 μl of λ concatamer ligation products (Chapter 7, *Protocol 9*) with 6 μl of BHB2688 extract and 2 μl of BHB2690 extract (*Protocol 6*).[b] If using a commercial extract, mix the amounts described in the manufacturer's protocol.

2 Incubate at 22 °C for 1 h.

3 Add 0.5 ml SM buffer plus one drop of chloroform to each packaging reaction. Place at 4 °C. Move quickly to the recombinant selection procedure (Section 3.4).

[a] See also Volume II, Chapter 3, *Protocol 13*.

[b] The extracts should be kept in a dry-ice bath until needed. The λ concatamer ligation products should contain up to 100 ng of DNA.

3 Plating out and recombinant selection

Whichever of the three methods is used to introduce recombinant molecules into *E. coli*, the number of cells that actually take up DNA will be relatively low. In *Protocol 3*, for example, DNA is added to about 10^8 cells but even under the best conditions only 10^7 transformants are expected. In most experiments, a proportion of these will not be recombinants but will contain self-ligated vector molecules. Even if the vector was phosphatase-treated to prevent self-ligation (Chapter 7, *Protocol 8*) a few molecules will have escaped the treatment and re-circularized without insert DNA.

The medium chosen for plating the cells should therefore be designed with two criteria in mind:

- Non-transformed cells should not be able to grow at all

- Recombinant colonies should be distinguishable from non-recombinant transformants

The precise strategy that is used depends on the genetic markers carried by the vector. As there are a large number of different vectors (Chapter 9) it might be expected that an equally large number of different selection strategies must be learned. Fortunately, this is not the case as most vectors are constructed along the same lines and only a few types of selection are used. The most important ones are:

(a) For plasmid vectors
  - antibiotic resistance
  - inactivation of β-galactosidase activity

(b) For M13 vectors
  - inactivation of β-galactosidase activity

(c) For λ vectors
  - selection on the basis of genome size
  - selection through inability to infect a P2 lysogenic strain of *E. coli*
  - inactivation of the *cI* gene
  - inactivation of β-galactosidase activity

3.1 Selection of plasmid vectors carrying antibiotic resistance genes

The first cloning vector to gain widespread use was pBR322 (Chapter 9, Section 3.2). pBR322 carries two genes, one that codes for a β-lactamase that provides resistance to ampicillin, and one (actually a set of genes) that codes for tetracycline resistance. When pBR322 is used as a cloning vector, the insert DNA is placed within one of these two genes, which one depending on the restriction site used. If, for instance, the *Bam*HI site is used then the tetracycline resistance gene is inactivated. The results of DNA uptake can therefore be determined by assessing the ampicillin and tetracycline responses of the individual cells: untransformed cells are amp^{s}tet^s, recombinant cells are amp^{r}tet^s, and cells transformed with self-ligated pBR322 are amp^{r}tet^r. The selection strategy therefore involves plating cells on to ampicillin agar, which screens out untransformed cells, and then transferring to tetracycline agar to distinguish the recombinants. The procedure, described in *Protocol 8*, can be adapted for identification of recombinants with any vector that uses antibiotic resistance as the selection method.

Protocol 8

Recombinant selection with pBR322

Equipment and Reagents

- LB agar plates (see *Protocol 6*). Supplement with filter-sterilized stock solutions of antibiotics (see *Appendix 2*) to give final concentrations of 40 μg/ml ampicillin or 15 μg/ml tetracycline

- LB medium (see *Protocol 1*)
- Equipment for replica plating, or sterile toothpicks

Method

1 Prepare and dry four LB-ampicillin plates, one LB-tetracycline plate, and one LB plate; also prepare an overnight culture of the host *E. coli* strain in 5 ml LB.

2 After expressing the cells (*Protocol 4*), make a dilution series in microfuge tubes:

Tube 1: 200 μl cells

Tube 2: 20 μl cells + 180 μl LB

Tube 3: 2 μl cells + 198 μl LB

3 Spread the contents of each tube on to an LB-ampicillin plate. For the spread-plate technique see Chapter 2, *Protocol 3*.

4 Prepare the following controls:

(a) Spread 200 μl of transformed cells on to an LB plate.

(b) Spread 200 μl of the overnight culture (untransformed cells) on to an LB-ampicillin plate.

5 Incubate all the plates overnight at 37 °C.

6 Check the results. The LB plate from step 4a should show confluent growth, indicating that the cells are viable. The untransformed cells should not grow on the LB-ampicillin plate (step 4b). The transformed cells plated on to LB-ampicillin should produce discrete colonies, derived from individual amp^r cells.

7 Choose the LB-ampicillin plate that shows the largest number of separated colonies and make a replica-plate on to LB-tetracycline. Alternatively, use sterile toothpicks to transfer small amounts of numbered colonies from the LB-ampicillin plate on to LB-tetracycline.

8 Incubate the LB-tetracycline plate at 37 °C overnight.

9 Colonies that grow on the LB-tetracycline plate are amp^r tet^r and so contain self-ligated pBR322 molecules. Colonies that do not grow are recombinants—amp^r tet^s. These can be recovered from the LB-ampicillin plate, and inoculated into broth for further study.

The procedure is straightforward but occasionally problems can arise.

(1) *No transformed colonies are obtained*. If no colonies appear on the LB-ampicillin plate then check the LB control (step 4a) to determine if the transformed bacteria are viable. If the LB plate shows confluent growth, as expected, then the most likely explanation is that DNA uptake has been unsuccessful, for example because of an error during the preparation of competent cells. The only other possibility is that too much ampicillin is present in the medium.

(2) *Confluent growth on the LB-ampicillin plates*. This is due to inactivation of the ampicillin when the plates are poured. Ampicillin is very sensitive to heat and can be substantially degraded if it is added to the molten agar before this has cooled down sufficiently. Make sure the instructions in *Appendix 2*, Section 1.3 are followed.

(3) *All the colonies from the LB-ampicillin plate grow on LB-tetracycline*. No recombinant molecules have been constructed. Check that you have correctly followed the protocols in Chapter 7.

(4) *Colonies on the LB-ampicillin plate are surrounded by haloes of smaller colonies.* The β-lactamase produced by amp^r bacteria is extracellular and diffusible, so will spread into the agar around resistant colonies. The local ampicillin concentration in the agar can therefore be depleted allowing nearby non-transformed cells (which are not killed by ampicillin, merely prevented from dividing) to grow and produce small colonies. The problem is particularly prevalent with high copy number vectors, presumably because these direct the synthesis of relatively large amounts of β-lactamase. An effective solution can be difficult to find. Try increasing the ampicillin concentration to 60 or 70 µg/ml and make sure that no inactivation is occurring during medium preparation. Do not incubate the plates for too long.

3.2 Lac selection of plasmids

Many plasmid cloning vectors employ a system called Lac selection, which centres around the plasmid-borne gene *lacZ'*, coding for the first 146 amino acids of β-galactosidase. This is the enzyme responsible for converting lactose to glucose plus galactose in the normal *E. coli* bacterium. The segment coded by *lacZ'* is not by itself sufficient to catalyse the conversion, but it can complement a host-encoded fragment to produce an active enzyme. Enzyme activity can be assayed with a chromogenic substrate, such as X-gal, which is colourless but is converted to a product with an intense blue product as a result of β-galactosidase activity. The assay is very sensitive and unambiguous.

A typical vector of this type is pUC18 (Chapter 9, Section 3.2). pUC18 also carries an ampicillin resistance gene, so transformants are plated on to ampicillin agar, on which all cells containing a vector molecule are able to grow to produce colonies. The cloned DNA is inserted into a restriction site within *lacZ'*, which means that recombinants are amp^r lacZ$^-$ and non-recombinants are amp^r lacZ$^+$. The two types of colony can be distinguished by including X-gal in the agar, as recombinants will be white and non-recombinants blue. Unlike pBR322, this system therefore allows recombinants to be identified during the first plating-out.

Protocol 9

Recombinant selection with pUC18

Reagents

- LB agar plates, including some containing ampicillin (see *Protocol 8*)
- LB medium (see *Protocol 1*)
- 2% (w/v) X-gal in dimethylformamide. It is best to prepare fresh X-gal for each experiment as solutions (and the solid) are light-sensitive. Stocks can, however, be stored with care at −20 °C in light-proof tubes. **Caution: Avoid exposure to dimethylformamide—follow the supplier's safety guidelines.**
- 100 mM IPTG.^{a} Stocks can be stored at −20 °C

Method

1. Prepare and dry four LB-ampicillin plates and one LB plate. Also prepare an overnight culture of the host *E. coli* strain in 5 ml LB.

2. After expressing the cells (*Protocol 4*), make a dilution series in microfuge tubes:

   Tube 1: 200 μl cells
   Tube 2: 20 μl cells + 180 μl LB
   Tube 3: 2 μl cells + 198 μl LB

3. Add 50 μl 2% (w/v) X-gal to each tube, along with 10 μl 100 mM IPTG.

4. Immediately spread the contents of each tube on to an LB-ampicillin plate. See Chapter 2, *Protocol 3* for the spread-plate technique.

5. Controls:[b]

   (a) Spread 200 μl of transformed cells on to an LB plate.

   (b) Spread 200 μl of the overnight culture (untransformed cells) on to an LB-ampicillin plate.

6. Incubate all the plates overnight at 37 °C.

7. The controls should show confluent growth of the transformed cells on LB and no growth of the untransformed cells on LB-ampicillin. The transformed cells should produce discrete colonies on LB-ampicillin, some of these colonies being blue (non-recombinants) and some white (recombinants).

[a] IPTG is a non-metabolizable inducer of the *lac* operon and is therefore needed to switch on expression of *lacZ'*.

[b] If you have used a phosphatased vector (Chapter 7, *Protocol 8*), and therefore expect no non-recombinants, then you will need an additional control to check that the colour reaction is working. Transform an aliquot of cells with 1 ng unrestricted pUC18 vector and treat in the same way as one of the test dilutions.

Recombinant selection with a *lacZ'* plasmid is subject to the same potential problems as pBR322 (see Section 3.1). The X-gal system also presents its own sources of error.

(1) *All the colonies are white.* If the colour reaction does not work then probably the X-gal has been degraded. Use a fresh stock and make sure the microfuge tubes are clean. Alternatively, the IPTG may be at fault, but this is less likely unless the stock is more than 3 months old or has been left to stand at room temperature.

(2) *The colour reaction is faint.* Some vectors do not produce a dense blue coloration. To enhance the reaction, leave the plates at 4 °C for 4–6 h after the colonies have grown. If the colour change is still ambiguous then suspect partial inactivation of the X-gal (see above). Some companies market related compounds that may give a more intense colour than X-gal.

(3) *The colonies are not uniformly coloured.* There are two common variants:

   (a) The periphery is more densely coloured than the centre: usually this is a non-recombinant.

   (b) The colony is white but there is a faint blue region in the centre: typically this is a recombinant.

An additional general problem with Lac selection is that it is not entirely trustworthy. The *lacZ'* gene is very accommodating and can function even though quite large pieces of DNA (up to 100 bp) are inserted into it, so long as the reading frame is maintained. This is one reason for the popularity of *lacZ'* vectors, as synthetic polylinkers carrying restriction sites for cloning purposes can be inserted into the gene without inactivating it. Unfortunately, it also means that some recombinants will still produce active β-galactosidase and so will appear blue on X-gal plates. If you are cloning small fragments of DNA, do not assume that blue colonies are uninteresting and use a second method (e.g. colony hybridization: Volume II, Chapter 5) to check them.

A second general problem is that occasionally a white colony will not contain inserted DNA. Sequence examination of the vector usually reveals a small deletion in the *lacZ'* gene, probably as a result of excision of the inserted DNA at some stage in colony growth. This is a general problem that can arise with any cloning vector, especially if the host strain is not chosen with care (see *Appendix 1*), or if the inserted DNA can form stem-loops that will enhance recombination and rearrangement events. However, with *lacZ'* vectors the problem can be worse because the presence of the polylinker stimulates stem-loop formation under some circumstances.

3.3 Recombinant selection with M13 vectors

All M13 vectors carry the *lacZ'* gene (Chapter 9, Section 5) and so recombinants are selected on X-gal plates in a manner similar to that described for pUC18. The plating-out procedure is different as the transformed cells will give rise not to colonies but to plaques, produced by M13 infection at localized positions on a lawn of bacteria. Each plaque represents a different transformant and its colour —clear or blue—indicates whether it is a recombinant or non-recombinant.

Usually, 1 ng of supercoiled M13 vector gives rise to about 10 000 blue plaques. After construction of recombinant molecules about 1000 plaques are expected per nanogram of DNA, although this assumes that all the original vector molecules were restricted. Since the transformation efficiency is so much greater with unrestricted vector a small proportion of unrestricted molecules will cause the resulting plates to be swamped with blue plaques. This makes it difficult to predict how much DNA should be used in the transformation experiment. A single plate should ideally have about 500 plaques in order to achieve good separation. Start by transforming with 1 ng of recombinant molecules but be prepared to increase or decrease this figure in subsequent experiments if necessary.

Protocol 10
Recombinant selection with M13mp18[a]

Equipment and reagents

- Water bath or incubator at 50 °C
- Exponential culture of an appropriate strain of E. coli[b]
- YT agar plates (8 g bacto-tryptone, 5 g bacto-yeast extract, 5 g NaCl, 15 g bacto-agar per 1 l). Check the pH and adjust to 7.0–7.2 with NaOH. Sterilize by autoclaving at 121 °C, 103.5 kPa (15 lb/in^2) for 20 min

- YTS agar (same as YT agar but with only 6 g bacto-agar/l)
- 2% (w/v) X-gal in dimethylformamide (see Protocol 9)
- 100 mM IPTG (see Protocol 9)

Method

1. Prepare and dry a YT plate. Melt 3 ml YTS agar in the microwave and place at 50°C.

2. To a tube of transformed cells add 50 μl 2% (w/v) X-gal, 10 μl 100 mM IPTG and 200 μl of exponential host E. coli cells.

3. Pour the contents of the tube into the molten YTS agar and pour on to the YT plate. See Chapter 2, Protocol 10 for the pour plate technique.

4. Allow the top agar to harden (about 5 min at room temperature) and then incubate the plate at 37 °C overnight. Plaques will be fully developed after about 10 h.

[a] M13 vectors do not carry antibiotic resistance genes so plasmid expression is not applicable. This protocol therefore follows directly from Protocol 3.

[b] If the competent cells are being made immediately before transformation, then take 50 μl of the exponential culture from Protocol 1, step 2, and inoculate into 5 ml LB medium. Incubate at 37 °C while you complete Protocols 1 and 3.

The various problems associated with Lac selection, discussed in Section 3.2 with regard to plasmid vectors, also apply to recombinant selection with M13. Further problems that may arise include the following.

(1) *The colour reaction is faint.* M13 plaques do not give such an intense blue coloration as colonies obtained with a vector such as pUC18. It should, nonetheless, be possible to distinguish blue and clear plaques unambiguously. As well as the possibilities described in Section 3.2, the type of agar medium used influences the apparent colour of the plaques. A rich medium such as DYT (Appendix 2) is ideal for growth of plaques but its relative opaqueness makes discrimination of plaque colour more difficult. Protocol 10 uses YT agar, which is less opaque than DYT. LB agar is a suitable alternative or as a last resort a solid base of pure agarose can be used.

(2) *The plates are smeary.* Whorls and other 'artistic' patterns are diagnostic of moist plates. Make sure they are dried thoroughly before use. Less dramatic

smearing results when the plates are moved before the top agar has hardened.

(3) *There are lumps on the plate.* This is due to the top agar not being completely molten. It will not solidify when held at 50 °C but neither will it melt if it is not already completely molten. Even small lumps will ruin the resulting plate.

(4) *The plaques are not uniformly distributed.* A common problem is that plaques appear only on one half of the plate. This is generally ascribed to the plates being on an uneven surface when the top agar is poured on, as can happen if the plates are stacked as they are poured.

3.4 Recombinant selection with λ vectors

The λ vectors are generally used for specialized purposes such as genomic and cDNA cloning. Descriptions of these procedures appear elsewhere in this book (Volume II, Chapter 2, Genomic cloning; Volume II, Chapter 3, cDNA cloning) along with protocols for recombinant selection with the appropriate λ vectors. In this Section, I present a summary of the strategies that can be employed, along with a procedure that illustrates the most popular of these strategies. This procedure follows on from the *in vitro* packaging reaction described in *Protocol 7*.

3.4.1 Selection on the basis of size

As with most viruses, the λ phage particle has a specific requirement for DNA molecules of a particular size. Only linear DNA molecules between 37 and 52 kb, and carrying *cos* sites at the termini, will be packaged into λ phage heads. This fact has been used in the construction of λ vectors, many of which produce molecules less than 37 kb long if ligation occurs without DNA insertion. Only molecules that contain an insert will be packaged, so only recombinant phages are produced.

Although size selection might appear to be foolproof it is rarely sufficient on its own for complete exclusion of non-recombinant molecules. This is because non-recombinant vector molecules must themselves be propagated in order to produce vector DNA, and so are constructed with a dispensable segment (the 'stuffer' fragment) between the two arms of the vector itself. The stuffer fragment is excised by restriction and the inserted DNA ligated in its place, between the arms. The problem that arises is that stuffer fragments may be present in the ligation so can reinsert themselves into the vector, producing non-recombinants that are appropriate sizes for packaging. This can be avoided by using ethanol precipitation to remove the stuffer fragment, or by digestion with additional restriction enzymes to produce subfragments of the stuffer that have termini incompatible with the vector arms (see Chapter 7, Section 3.1). Even with rigorous technique, complete removal is not usually possible and it is normal for λ vectors to use a second type of selection to back up the size system.

3.4.2 Selection using the Spi phenotype

Vectors such as λEMBL3 and λEMBL4 (Chapter 9, Section 4.2) employ a system called Spi selection. This is based on the fact that wild-type λ phage are unable to infect a bacterium which already carries a P2 prophage—they are sensitive to P2 interference). The Spi phenotype is due to the activity of the *red* and *gam* genes on the λ genome, and if these genes are deleted the phage becomes Spi⁻ and able to infect a P2 lysogen of a suitable host strain. Vectors that use the Spi system carry the *red* and *gam* genes on the stuffer fragment so that non-recombinants are Spi⁺ and recombinants Spi⁻.

3.4.3 Insertional inactivation of the λ*cI* gene

Not all λ vectors are replacement vectors with dispensable stuffer fragments. Some are analogous to plasmid and M13 vectors, carrying restriction sites that allow insertion of new DNA and inactivation of a selectable gene. The vector λgt10 (Chapter 9, Section 4.2), as well as a few others, carry cloning sites within the *cI* gene, which is part of the immunity region and codes for one of the regulatory proteins that control the λ infection cycle. *cI*⁺ phages are highly efficient at forming lysogens in certain permissive host strains and as a result relatively few phage particles are produced. In contrast *cI*⁻ phages are inefficient at lysogeny and larger amounts of phage particles are produced. The *cI* genotype can be assessed by eye as *cI*⁺ plaques are turbid because they contain large numbers of intact lysogenic bacteria, whereas *cI*⁻ plaques are clear. Alternatively if a broth culture of a suitable host is infected with packaged phage, the bulk of the phage particles that are produced will be *cI*⁻, as the *cI*⁺ phage will form lysogens and be 'trapped' inside the cells.

3.4.4 Selection using *lacZ'*

Several λ vectors carry the *lacZ'* gene and recombinants are identified by the blue-clear colour reaction on X-gal agar, as described for M13 vectors (Section 3.3). In λgt11 and λZAPII (Chapter 9, Section 4.2), as well as a number of other λ vectors, the *lacZ'* gene carries restriction sites that allow insertional inactivation in the standard way. A few vectors carry the gene on the stuffer fragment, so that recombinants are *lacZ'*⁻ owing to complete loss of the gene.

Protocol 11 illustrates the general principles of recombinant selection with λ vectors, using λZAPII as the example. The starting point for this procedure is packaged λZAPII recombinants, obtained as described in *Protocol 7*.

4 Vectors combining features of both plasmids and phages

In recent years, the development of novel vector types has blurred the distinction between plasmid- and bacteriophage-based systems. Two classes of vectors, cosmids and phagemids, combine features of both plasmids and phage chromosomes and have to be handled in special ways.

Protocol 11

Recombinant selection with λZAPII[a]

Reagents

- *E. coli* XL1-Blue MRF′
- LB agar plates containing 12.5 μg/ml tetracycline (see *Protocol 8*)
- LB medium containing 10 mM MgSO$_4$ and 0.2% (w/v) maltose. Prepare LB medium as described in *Protocol 1*. Supplement immediately before use from a 1 M stock solution of MgSO$_4$, sterilized by autoclaving, and a 20% (w/v) stock solution of maltose sterilized by filtration
- SM buffer (see *Protocol 7*)

- NZY agar plates (10 g NZ amine, 5 g bacto-yeast extract, 2 g MgSO$_4$, 15 g bacto-agar per 1 l). Check the pH and adjust to 7.0–7.2 with NaOH. Sterilize by autoclaving at 121 °C, 103.5 kPa (15 lb/in^2) for 20 min
- NZY top agar (same as NZY agar but with 7 g agarose/l instead of the bacto-agar)
- 2% (w/v) X-gal in dimethylformamide (see *Protocol 9*)
- 500 mM IPTG (see *Protocol 9*)

Method

1 Streak a colony of *E. coli* XL1-Blue MRF′ onto an LB-tetracycline plate (see Chapter 2, *Protocol 2*) and incubate overnight at 37 °C.[b]

2 Inoculate 50 ml LB medium containing 10 mM MgSO$_4$ and 0.2% (w/v) maltose with a single colony from the streak plate. Incubate overnight at 30 °C with shaking.

3 Pellet the cells at 2000 r.p.m. for 10 min at 4 °C in a bench-top centrifuge and resuspend in 25 ml ice-cold 10 mM MgSO$_4$. Add additional 10 mM MgSO$_4$ to bring the OD at 600 nm to 0.5.

4 Place 200 μl of resuspended bacteria in each of three microfuge tubes. To the first tube add 1 μl of packaged λZAPII recombinants (prepared as described in *Protocol 7*). To the other two tubes add 1 μl of a 1:10 and 1:100 dilution of the packaged λZAPII recombinants, respectively.

5 Incubate at 37 °C for 15 min to allow the phages to adhere to the bacterial surfaces.

6 Prepare and dry three NZY plates. Melt 3 × 3 ml NZY top agar in the microwave and place at 50 °C.

7 To each tube of infected cells add 300 μl 2% (w/v) X-gal and 15 μl 500 mM IPTG. Add to 3 ml molten top agar and pour on to a NZY plate. See Chapter 2, *Protocol 10* for the pour plate technique.

8 Allow the top agar to harden (about 10 min at room temperature) and then incubate the plates at 37 °C overnight.

[a] See also Volume II, Chapter 3, *Protocol 14*.

[b] The tetracycline selection used when growing this host strain ensures that the F′ plasmid that carries the *lac* genes (as well as the tetracycline resistance gene) is not lost from the cells.

4.1 Cosmids

A cosmid (9) is essentially a plasmid that carries λ *cos* sites and is packageable into λ phage heads (Chapter 9, Section 3.6). Their advantage is that because they lack virtually all of the λ genome they can accommodate very large pieces of insert DNA, up to 45 kb or more, which is more than can be handled by an orthodox λ vector.

The procedures for dealing with cosmids are identical to those used for ordinary λ vectors. Ligation is carried out with a high DNA concentration to produce concatamers (Chapter 7, *Protocol 9*) and the products are inserted into phage heads by *in vitro* packaging (*Protocol 7*). The difference is that after infection of host *E. coli* cells a cosmid vector or recombinant molecule behaves like a plasmid, so a colony is produced, usually on an antibiotic plate, possibly utilizing Lac selection for recombinant identification. New 'phage' particles are not produced because the cosmid does not carry any of the standard λ genes. Recombinant selection is therefore carried out as described in *Protocols 8* and *9*, with the composition of the agar medium on to which the infected cells are plated determined by the particular features of the cosmid that is being used.

The main application of cosmid vectors is in the construction of genomic libraries, where the large insert capacity is an advantage in minimizing the number of clones required for a complete library. Procedures for using a cosmid vector when preparing a genomic library are given in Volume II, Chapter 2.

4.2 Phagemids

A phagemid combines features of plasmid and M13 vectors (10). The main use of an M13 vector is in production of single-stranded DNA, which is required for several procedures, notably DNA sequencing (Volume II, Chapter 6). Unfortunately, the M13 genome cannot be modified to any great extent without impairing its basic genetic functions, so M13 vectors are not themselves very flexible. In addition, they have quite small size capacities and are inefficient at cloning fragments of more than 3 kb.

Phagemids provide an alternative means of obtaining single-stranded DNA (Chapter 9, Section 3.3). They carry two replication origins, a standard plasmid origin and one derived from M13 or a related phage such as fl. The phage origin is the key component in the synthesis of single-stranded DNA, although this also requires enzymes and coat proteins coded by phage genes, which the phagemid lacks.

A phagemid can be treated in exactly the same way as a standard plasmid vector, using the transformation and selection strategies described in Sections 2.1, 2.2, 3.1, and 3.2. Alternatively, cells containing a phagemid vector can be 'super-infected' with a 'helper phage', which itself contains a modified genome, but one that retains the genes for single-stranded DNA production. The helper phage is able to convert the phagemids into single-stranded DNA molecules, which are assembled into defective phage particles and secreted from the cell.

When using a phagemid for single-stranded DNA production it is important to use both an appropriate host and the correct helper phage. The technique varies slightly depending on the combination that is required. *Protocol 12* provides details for the phagemid pUC118, which is generally cloned in *E. coli* MV1184 and superinfected with the helper phage M13KO7. If using one of the many phagemids that are commercially available then simply adjust the procedure in accordance with the supplier's instructions.

Protocol 12

Cloning with pUC118

Reagents

- M13KO7 phage stock. Phages are usually maintained as plaques on a lawn of *E. coli* MV1184 cells
- DYT medium, with and without 70 μg/ml kanamycin. Prepare DYT medium (16 g bacto-tryptone, 10 g bacto-yeast extract, 5 g NaCl per 1 l), check the pH and adjust

to 7.0–7.2 with NaOH; sterilize by autoclaving at 121 °C, 103.5 kPa (15 lb in^2) for 20 min. Immediately before use, supplement with a filter-sterilized stock solution of kanamycin (see *Appendix 2*)

- 35 mg/ml kanamycin solution, sterilized by filtration

A. Preparation of a phage stock for superinfection

1   Transfer a single isolated M13KO7 plaque into 2 ml of DYT + 70 μg/ml kanamycin. Incubate with moderate agitation overnight at 37 °C.

2   Centrifuge the culture at 10 000 *g* for 5 min to pellet the bacteria and store the phage supernatant at 4 °C.

3   Measure the phage titre (Chapter 2, Section 5.3), which should be greater than 10^{11}/ml.

B. Cloning procedure

1   pUC118 is derived from pUC18 and so carries the *lacZ'* gene as well as the gene for ampicillin resistance. Select recombinants as described in *Protocol 9*.

2   To produce single-stranded DNA versions of the recombinant phagemids, inoculate a single recombinant colony in 5 ml DYT.

3   Add 10^8 plaque forming units of M13KO7, prepared as described in part A.

4   Incubate at 37 °C for 1.5 h. Strong agitation is needed to obtain good yields of phage particles. The best procedure is to use test-tubes angled at 45° on a platform shaker, so that the cultures are vigorously aerated as well as being shaken.

5   Add 10 μl of 35 mg/ml kanamycin solution. Continue the incubation overnight. M13KO7 carries a kanamycin resistance gene so this treatment selects for super-infected cells.

6   Harvest extracellular 'phage' and prepare single-stranded DNA as described in Volume II, Chapter 6.

References

1. Smith, H. O., Danner, D. B., and Deich, R. A. (1981). *Ann. Rev. Biochem.* **50**, 41.
2. Mandel, M. and Higa, A. (1970). *J. Mol. Biol.* **53**, 159.
3. Cohen, S. N., Chang, A. C. Y., and Hsu, L. (1972). *Proc. Natl. Acad Sci. USA* **69**, 2110.
4. Oishi, M. and Cosloy, S. D. (1972). *Biochem. Biophys. Res. Comm.* **49**, 1568.
5. Hanahan, D. (1983). *J. Mol. Biol.* **166**, 557.
6. Shigekawa, K. and Dover, W. J. (1988). *Biotechniques* **6**, 742.
7. Scherer, G., Telford, J., Baidari, C., and Pirrotta, V. (1981). *Dev. Biol.* **86**, 438.
8. Rosenberg, S. M. (1987). In *Methods in enzymology* (ed. R. Wu and L. Grossman), Vol. 153, p. 95. Academic Press, New York.
9. Collins, J. and Hohn, B. (1978). *Proc. Natl. Acad. Sci. USA* **75**, 4242.
10. Dente, L., Cesareni, G., and Cortese, R. (1983). *Nucl. Acids Res.* **11**, 1645.

Chapter 9

Survey of cloning vectors for *Escherichia coli*

T. A. Brown

Department of Biomolecular Sciences, University of Manchester Institute of Science and Technology, Manchester M60 IQD, UK

1 Introduction

The vector plays the central role in a DNA cloning experiment and must be carefully chosen so that its properties are appropriate for the objectives of the project. The features of the vector must then be taken into account when designing the experimental strategy that will be used to generate and identify recombinant clones. An understanding of the different types of vector that are available, and the properties of each one, is therefore a prerequisite for successful molecular biology.

Over the years, an almost bewildering number of different vectors have been designed. Comprehensive details of 150 vectors in current use have recently been published (1) and similar information on a much larger set of older vectors is also available (2). In this chapter I will not attempt to reproduce the information in these compendia but will instead describe the general features of the different classes of cloning vector, with specific details limited to selected examples of each type. The chapter is divided into three sections, which reflecting the three major subdivisions of cloning vector. These are:

- Plasmid-based vectors
- Vectors based on bacteriophage λ
- Vectors based on bacteriophage M13

Although these three types of cloning vector are quite different from one another, they all have one property in common: each is designed for use with *Escherichia coli* as the host organism. Before examining the vectors we should, therefore, understand why the vast majority of cloning experiments are carried out with this bacterium.

2 *E. coli* as the host organism for recombinant DNA research

2.1 Advantages of *E. coli* as a host organism

There are several reasons why most people use *E. coli* as a host organism for gene cloning experiments, even if the vector that is employed is intended eventually to be used in another organism such as yeast. Much knowledge has been gathered during the past decades with regard to the structure and properties of bacterial plasmids and bacteriophage DNA molecules, and the genetics and biochemistry of plasmid replication and transfer and the bacteriophage infection cycle are well understood, including important issues such as the control of gene expression during these events.

Genetic studies with plasmids from Gram-negative bacteria have identified regions essential for replication of the plasmid, for transfer of plasmids between bacteria by means of conjugation, and for mobilization of plasmids. In the latter process the concerted action of the functions from more than one plasmid is required. Mobilization of a plasmid from one bacterium to the other requires the presence, in the bacterium containing the mobilizable plasmid, of a second plasmid providing in *trans* the functions for transfer of the plasmid (3). In addition, functions (e.g. *par*) (4) have been identified that control the distribution of plasmids to daughter cells at the time of cell division. Other functions have been identified in regions of plasmid DNA that control site-specific recombination (e.g. *cer*) (5), a process implicated, through the interconversion of monomeric and multimeric forms of plasmid DNA, in plasmid stability, and in a region variably called *rom* or *rop* (repressor of primer) (6) which controls plasmid copy number.

The mechanisms that control gene expression in *E. coli* have been extensively studied and are, generally speaking, well understood. Gene expression is largely controlled at the level of initiation of transcription. For a large number of genes and operons the sites at which initiation of transcription takes place have been identified and their nucleotide sequences have been determined. In a number of cases, the control of gene expression is also exerted at the level of transcription termination and/or anti-termination. In addition, regulation of gene expression takes place at the level of RNA processing and RNA degradation, which are intimately coupled to translation. Last but not least, expression of some genes is also controlled during initiation and elongation of the translation product. For a number of genes, such as those of the lactose (*lac*) operon, the tryptophan (*trp*) operon and the genes of the leftward (p_L) and rightward (p_R) operons of bacteriophage λ, we have detailed knowledge of the nucleotide sequences pertaining to these control mechanisms, the identities and structures of the regulatory proteins, and the nature of the interaction between the nucleotide sequences and the regulatory proteins (7). Expression from these promoters can be modulated by negative or positive control elements, such as repressors and activators, and the structure and function of these elements is known in great detail (8). This knowledge is being exploited to try to achieve efficient expression of genes in *E. coli* and to fine-tune the levels of expression.

Other advantages of using *E. coli* as a host organism are that a great variety of well-characterized laboratory strains with useful features are available and the fact that such strains can, in general, be cultivated in well-defined media. Of the many *E. coli* strains that exist, several are frequently used for cloning and expression of genes, sometimes because such strains have proven to yield high transformation frequencies, sometimes because they contain genetic characteristics that are required when a particular vector is used, and in other instances because expression and/or stability of a gene product is superior in one strain compared with others. A list of *E. coli* strains that are frequently used in gene cloning, together with a description of their genetic markers, is given in *Appendix 1*.

Finally, it is also worth mentioning that, in most countries, recombinant DNA research with *E. coli* is classified in the low-risk category and is, in many cases, exempt from regulations. This means that the great majority of experiments with *E. coli* can be carried out under standard conditions (see Chapter 2) and so do not require the more expensive containment facilities needed for research with more dangerous microbes.

2.2 *E. coli* is not an ideal host for gene cloning

Although *E. coli* has many attractive features for cloning and expression of genes, it is not an ideal or universal host organism and there is a limit to its effectiveness, especially in the synthesis of proteins specified by cloned genes (*Table 1*). In particular, the use of *E. coli* for the industrial production of pharmaceutical proteins and food additives is hindered by the presence of an endotoxin which may diminish general public acceptability. This assumes that the protein products can be synthesized in *E. coli* in the first place and, in many cases, they cannot—at least not efficiently, because *E. coli* is unable to perform a number of specific functions that are needed for the expression of typical genes from eukaryotic organisms. *E. coli* cannot remove introns from the transcripts of discontinuous eukaryotic genes, which is why the cloning of DNA copies of eukaryotic RNA (cDNA cloning: see Volume II, Chapter 3) is so important in molecular biology.

Table 1 Advantages and disadvantages of *Escherichia coli* as a host of gene cloning

Advantages	Disadvantages
Genetics and biochemistry are well-understood	Not a GRAS (generally regarded as safe) organism
Plasmids are stable	No capacity for intron splicing
Many types of vector	Limited secretion of proteins into the culture medium
Many strains with useful mutations	Proteins are not glycosylated
High yield of vector DNA	Does not synthesize disulphydryl bonds
High yield of protein	Proteins might be insoluble
	Proteins might be unstable

Less easy to circumvent is the inability of *E. coli* to process its primary translation products, which means that proteins expressed from eukaryotic genes cloned in this host do not display the post-translational chemical modifications that frequently are essential for full activity and stability of the protein. For example, *E. coli* is unable to carry out *N*- or *O*-linked glycosylations, to completely remove the *N*-terminal methionine residue from a number of proteins, or to acetylate, phosphorylate or palmitate proteins. Although for a number of proteins it has been reported that the absence of, for example, *N*-linked glycosyl residues, does not effect the biological activity, the lack of these modifications means that the immunological properties of the product are different from those of the natural protein, and the protein may have a reduced stability or reduced retention time if used as a pharmaceutical. Because *E. coli* has an internal reducing environment, proteins that in their active conformation contain disulphydryl bridges cannot fold properly and often will be deposited as biologically inactive inclusion bodies inside the bacterium. If any of the processing events described above are a prerequisite for the proper functioning of the protein product, then *E. coli* can be ruled out as a host organism.

3 Plasmid vectors

Plasmid-based vectors can be divided into eight categories (1):

(1) *Basic plasmid vectors* are the simplest type, being little more than an origin of replication attached to one or more selectable markers. They are the workhorses of molecular biology and have applications in many of the procedures described in Volume II, in particular being used to clone short DNA fragments that will be sequenced by the double-stranded method (Volume II, Chapter 6) or used as hybridization probes (Volume II, Chapter 5). Examples are the pUC series of vectors.

(2) *Phagemids* have both a plasmid origin of replication and an origin from a phage with a single-stranded DNA genome, such as M13 or f1. The phage origin enables the cloned sequence to be obtained as single-stranded DNA for use in procedures such as DNA sequencing (Volume II, Chapter 6). The pEMBL vectors are examples of this type.

(3) *RNA expression vectors* carry one or more promoter sequences and so can be used for *in vitro* synthesis of RNA transcripts of cloned DNA fragments, which is particularly useful in the production of hybridization probes (Volume II, Chapter 5). The pSP vectors marketed by Promega are simple examples.

(4) *Protein expression vectors* are designed in such a way that the cloned gene can direct synthesis of recombinant protein inside the *E. coli* host cell (Volume II, Chapter 8). There are many examples, such as pKK223-3 and the pGEX series (both available from Pharmacia), and the pET series.

(5) *Cosmids* combine features of plasmid and λ-based vectors and are used to clone large pieces of DNA when genomic libraries are prepared (Volume II, Chapter 2). Examples are pJB8 and the pcosEMBL vectors.

(6) *Cloning vectors for PCR products* have special features that enable them to clone DNA fragments generated by PCR amplification (Volume II, Chapter 7). Several commercial ones are available, including pGEM-T (Promega), the pCR series (Stratagene) and pDIRECT (Clontech).

(7) *Promoter probe vectors* are a more specialized type that are used to identify DNA fragments that contain promoter sequences; e.g. pKK175-6.

(8) *Mutagenesis vectors* are designed for use in *in vitro* mutagenesis experiments; e.g. the pALTER series (Promega).

Note that some vectors combine the features of two or more of the types listed above. For example, popular commercial vectors such as pBluescript (Stratagene) and some of the pGEM series (Promega) are phagemids/RNA expression vectors and the pGEMEX vectors (Promega) are phagemids that allow expression of both RNA and protein. These vectors can also be used quite effectively for general methods such as production of clones for double-stranded DNA sequencing and hybridization probing.

The following sections describe the general properties of plasmid cloning vectors and the specific features of examples of the first six categories listed above. Promoter probe and mutagenesis vectors are omitted because these types have specialized applications that are beyond the scope of this book.

3.1 General properties of plasmid vectors

3.1.1 Plasmid copy number and its control

Most plasmid-based vectors currently used in recombinant DNA experiments are derivatives of the naturally occurring plasmid ColEl or the ColEl-related plasmid pMB1. A few are derived from p15A, pSC101 or R6K. The important component of the original parent plasmid that is retained by the cloning vector is the origin of replication, a nucleotide sequence a few hundred base pairs in length, which determines the number of copies of the vector that are present in each transformed bacterium. Each of the five parent plasmids has a copy number of 10–20, and vectors based on p15A, pSC101 and R6K have similar copy numbers to their parents. However, all commonly used vectors with a ColE1 or pMB1 origin also carry mutations that increase the content to up to 80 per cell, and this number can be further increased, perhaps to > 1000 per cell, by incubation of the transformed culture with an inhibitor of protein synthesis such as chloramphenicol. This procedure, which is called plasmid amplification (*Protocol 1*), works because ColE1 and pMB1 are 'relaxed' plasmids, which means that their replication origins can function in the absence of protein synthesis. This contrasts with the origin of replication for the bacterial chromosome which is inactive if protein synthesis is inhibited (as are the origins of 'stringent' plasmids such as p15A, pSC101 and R6K). Plasmid replication therefore continues in the absence of chromosomal DNA replication and bacterial cell division. The increased plasmid content results in substantially higher yields of plasmid DNA after extraction by a large-scale method (see Chapter 3, *Protocol 5*).

Protocol 1

Plasmid amplification

Reagents

- Culture of E. coli transformed with a plasmid vector with a ColE1 or pMB1 origin of replication
- Chloramphenicol

Method

1   Incubate the transformed E. coli culture at 37 °C until an OD at 550 nm of 0.8–1.0 is reached.

2   Add solid chloramphenicol to a final concentration of 150 µg/ml.

3   Continue to incubate at 37 °C overnight.

3.1.2 Fragment size, compatibility and mobilization

The size of DNA fragments that can be cloned in plasmid vectors is virtually un-limited, but the stability of very large recombinant plasmids may be diminished. In addition, plasmid copy number and the transformation capacity of plasmids are also unfavourably affected by the insertion of very large DNA segments. In practice, a plasmid vector is rarely used to clone DNA fragments of > 10 kb.

Plasmids that are closely related usually cannot be maintained together within the same bacterium. For example, plasmids derived from ColE1 or pMB1 are 'in-compatible' and their derivatives cannot coexist within the same bacterium. Only a plasmid that belongs to a different compatibility group can be main-tained within a bacterium together with ColE1-related plasmids. If one wishes to introduce into the same bacterium two genes or regulatory elements present on different plasmids, one of which has a ColE1-like origin of replication, then the second plasmid must be derived from p15A, pSC101, R6K or some other parent.

Most plasmid vectors that are currently used are non-conjugative because they lack genes for mobilization. This means that genetic material can be transferred to another bacterium only if the cell contains additional plasmids providing the mobilization and conjugation functions. A second requirement for transfer of genetic material is the presence in the plasmid vector of an element called *bom*, for *basis of mobilization*. In some vectors, this element has been deleted to limit further the adventitious spread of cloned sequences.

3.1.3 Plasmid selection

Because most transformation procedures are relatively inefficient (under optimal conditions only a few per cent of all bacteria are transformed) it is essential that one can discriminate transformed and non-transformed bacteria (see Chapter 8, Section 3). With most vectors this is accomplished by making use of dominant selection markers, such as genes carrying an antibiotic resistance marker which, if present on the plasmid, provide the bacterium with resistance to that antibiotic. The most commonly used marker is that for ampicillin resistance (*amp*) originating from transposon Tn3. Other markers that are currently in use include the tetracycline (*tet*) gene from Tn*10*, the kanamycin resistance (*kan*) gene from Tn*903* and the neomycin resistance (*neo*) gene from Tn5. All of these markers are similar in that direct selection for plasmids containing one of the genes is easy and efficient. For example, *E. coli* bacteria are sensitive to very low concentrations of ampicillin (a few μg/ml), but become resistant to very high levels (up to 1 mg/ml) if the *amp* gene is expressed.

The earliest cloning vectors employed a selection system based on two dominant selection markers. The rationale was that if a DNA fragment is inserted into one of these markers, selection can be based on the other marker and inactivation of the first marker is evidence for the successful cloning of the DNA fragment (see Chapter 8, Section 3.1). This procedure is unwieldy because it requires transfer of clones from one antibiotic medium to another in order to assess fully the genotypes and identify recombinants. Most modern vectors employ a different method, called direct selection, which enables transformant selection and recombinant identification to be accomplished on a single agar plate. In this procedure, transformants are selected by virtue of expression of an antibiotic resistance gene carried by the vector, and recombinants are identified because in these a second marker, one whose expression can be directly assayed, is inactivated. The commonest system is Lac selection (see Chapter 8, Section 3.2), which requires a vector that carries the *lacZ'* gene, a shortened version of the *E. coli* gene coding for the enzyme β-galactosidase. In an appropriate host strain (see *Appendix 1*), a plasmid with a *lacZ'* gene enables its host bacterium to split lactose and related compounds into their monosaccharide components. If the chromogenic lactose analogue X-gal is included in an agar medium, along with an inducer of the lactose operon, such as IPTG, transformed colonies are coloured blue. Inactivation of *lacZ'*, through insertion of cloned DNA at the appropriate site in the vector, yields a white colony.

3.2 Basic plasmid vectors

The earliest cloning vectors to be constructed comprised little more than a plasmid origin of replication linked to two or more selectable markers. One or both of these markers contain recognition sequences for restriction endonucleases that cut the vector at just that single position and hence can be used to linearize the plasmid prior to ligation in the presence of the DNA fragments to be cloned. Although these basic vectors lack the additional nucleotide sequences needed for more sophisticated applications, such as generation of single-stranded versions of the cloned DNA, or efficient synthesis of RNA and protein, their small size and relative simplicity makes them ideal for the job they were designed for, which is to clone short (less than 10 kb) fragments of DNA. Several of these vectors are still in use, notably pBR322 and its relatives, and the pUC series of direct selection vectors.

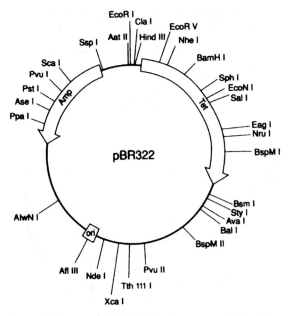

Figure 1 Structure of the basic cloning vector pBR322. The vector is 4363 bp and carries the pMB1 origin of replication and the *amp* and *tet* markers. Restriction sites that can be used for cloning are *Bam*HI, *Bsp*MI, *Eco*31I, *Eco*NI, *Eco*RV, *Nhe*I, *Nru*I, *Pst*I, *Pvu*I, *Sal*I, *Sca*I, *Sph*I and *Xma*III. Reproduced with the kind permission of Amersham Pharmacia Biotech AB.

3.2.1 pBR322 and its relatives

pBR322 (*Figure 1*) is 4363 bp and carries the *amp* and *tet* selectable markers (9). For many years it was the most widely-used cloning vector, despite the need to test for both ampicillin and tetracycline resistance when searching for recombinant clones (see Chapter 8, Section 3.1). Its popularity was due largely to the range of unique restriction sites present in the *amp* or *tet* genes, enabling fragments with any of a variety of sticky ends to be ligated into the vector. It is still occasionally used but is not generally recommended because of three disadvantages inherent in its construction:

(1) It lacks the *cer* site, which means that plasmid multimers are easily formed during replication (10). This is a disadvantage because it can lead to the loss of plasmids during cell division. The critical point is that the copy number of a plasmid refers to the number of plasmid copies, not the number of molecule copies. pBR322 has a copy number of approximately 80, which corresponds to just five or six molecules if each of these is a multimer comprising 14–16 linked plasmid copies. With such a small number of molecules per cell there is a relatively high chance of daughter cells being produced that have no plasmid copies. These plasmid-free cells will not display antibiotic resistance so under most circumstances will be lost; however, in the absence of selection they will become predominant in the culture as they will have a slight

growth advantage because of the reduced amount of DNA synthesis that they need to carry out per cell division. This can become a significant problem during storage of clones on agar plates, even when the plates contain antibiotic, because the localized concentration of antibiotic around the colonies decreases over time, owing to the secretion of inactivating enzymes by the resistant bacteria, thereby allowing growth of untransformed cells. The absence of the *cer* site is a feature of pBR322 and all of its relatives.

(2) Owing to the presence of the *bom* site, pBR322 can be mobilized by the presence of conjugative plasmids, such as ColK, which deliver in *trans* the mobilization functions. This raises the possibility that recombinant pBR322 molecules could escape from their host cells, for example after accidental ingestion. There is only a remote possibility that this will happen but it precludes the use of pBR322 in some cloning projects (refer to your local regulations for guidance). Plasmid pBR327 (11) is a deleted version of pBR322 which lacks the *bom* site and hence is non-mobilizable.

(3) pBR322 also lacks a partition function needed for proper segregation of plasmid vectors to daughter cells during cell division, again leading to plasmid-free cells in the absence of selection, as described in point 1. To overcome this segregational instability, a derivative of pBR327 has been constructed, called pBR327*par* (12), containing the *par* locus from pSC101. Except for its improved stability properties, pBR327*par* has the same characteristics as pBR327.

3.2.2 The pUC series of plasmid cloning vectors

Plasmids of the pUC series (13, 14) are the most popular of the basic plasmid vectors. They are direct selection vectors that contain the *lacZ'* gene and are plated onto agar medium containing ampicillin, X-gal and IPTG in order to identify recombinants (see Chapter 8, Section 3.2). The *lacZ'* gene contains the promoter–operator region of the lactose operon and the first 146 codons of *lacZ*, the *E. coli* gene coding for β-galactosidase. This part is sufficient to complement a deleted version of *lacZ* (called *lacZΔM15*) carried by the host bacterium. The *lacZ'* gene on the vector contains an inserted DNA fragment carrying a series of restriction sites used for cloning purposes. Although this 'polylinker' disrupts the coding region of *lacZ'* it does not affect the activity of the protein product, largely because the insertion is designed so that it contains a series of codons that fuse in-frame with the gene, allowing read-through of the transcript by the ribosome.

Most of the vectors of the pUC series fall into pairs, each member of the pair having the same polylinker but in opposite orientations (*Figure 2*). This is particularly useful because it provides a convenient means for 'turning around' a fragment that has been cloned into one of the internal sites within the polylinker. This is achieved by excising the cloned fragment by cutting at the *Eco*RI and *Hin*dIII sites that flank the polylinker and then religating the excised fragment into the other member of the vector pair, after this has itself been restricted with *Eco*RI and *Hin*dIII. This manipulation is relevant if the design of the experiment demands that the cloned fragment has a particular orientation within the vector.

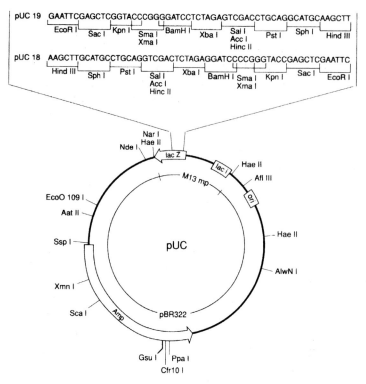

Figure 2 Structures of the basic cloning vectors pUC18 and 19. These vectors are 2686 bp and carry the pMB1 origin of replication and the *amp* and *lacZ'* markers. The restriction sites that can be used for cloning are *Acc*I, *Bam*HI, *Eco*RI, *Hinc*II, *Hind*III, *Kpn*I, *Pst*I, *Sal*I, *Sma*I, *Sph*I, *Sst*I, *Xba*I and *Xma*I. Other vectors in the pUC series have different cloning sites as follows. pUC7: *Acc*I, *Bam*HI, *Eco*RI, *Hinc*II, *Pst*I and *Sal*I. pUC8 and pUC9: *Acc*I, *Bam*HI, *Eco*RI, *Hinc*II, *Hind*III, *Pst*I, *Sal*I, *Sma*I and *Xma*I. pUC12 and pUC13: *Acc*I, *Bam*HI, *Eco*RI, *Hinc*II, *Hind*III, *Pst*I, *Sal*I, *Sma*I, *Sst*I, *Xba*I and *Xma*I. Reproduced with the kind permission of Amersham Pharmacia Biotech AB.

pUC vectors have a very high copy number (500–700 plasmids per cell) largely because they lack a functional *rop* gene (as is the case with pBR327), but also because of a fortuitous mutation in the region containing the origin of replication that occurred early during construction of the pUC series. This mutation results on a shortened form of the transcript called RNA-I, which plays a role in copy number regulation. This high copy number means that large yields of vector DNA can be obtained from cultures transformed with a pUC plasmid, making these vectors ideal for applications such as the preparation of hybridization probes (Volume II, Chapter 5) and DNA sequencing (Volume II, Chapter 6).

3.3 Phagemids

The standard method for DNA sequencing used throughout the 1980s and much of the 1990s required that the DNA to be sequenced was available in single-stranded form. Originally, M13 vectors were used to obtain single-stranded DNA, but vectors based on this bacteriophage can only be used to clone fragments shorter than 3 kb; recombinant molecules containing larger inserts tend to be unstable (see Section 5). Phagemids were developed to overcome this problem and enable single-stranded versions of longer cloned fragments to be obtained. Although sequencing technology has now moved on and single-stranded methods have been largely superseded by double-stranded sequencing and procedures based on thermal cycling (Volume II, Chapter 6), phagemids are still required for specialized techniques such as *in vitro* mutagenesis and many of the commercial multipurpose vectors (Section 3.8) are phagemids.

The key feature of a phagemid is the presence of two origins of replication, one derived from a plasmid and directing double-stranded replication of the vector, and a second derived from a filamentous bacteriophage with a single-stranded DNA genome (e.g. M13 or f1). This origin can be activated by super-infection of a transformed culture with a helper phage, such as M13KO7 (15), which provides in *trans* the genes for the phage replication enzymes and coat proteins that are needed for single-stranded DNA propagation (see Chapter 8, *Protocol 12*). The phage particles that are extruded into the medium contain either single-stranded helper phage DNA or single-stranded plasmid DNA (16). The amounts of single-stranded phage DNA and plasmid DNA produced are, in general, considerably lower than those obtained with an M13 vector, probably because of interference between plasmid replication and phage multiplication. The relative amounts of the single-stranded products—phage DNA and plasmid DNA—may also vary and depend both on the type of plasmid and helper phage used. To optimize the yield of single-stranded plasmid DNA, phage mutants have been isolated that show interference resistance. These mutants can increase the yield of single-stranded plasmid DNA by tenfold.

Most phagemids come in pairs, the two vectors differing in the orientation of the origin of replication of the filamentous phage. The orientation of the origin determines which of the two strands of the plasmid will be encapsulated in the phage particle. In the terminology that is used, '+' indicates that the polarity of the origin of plasmid and phage replication is the same, and '−' signifies that the two origins have the opposite polarity.

3.3.1 The pEMBL phagemids

pEMBL8 and pEMBL9 (*Figure 3*) (17) are a pair of phagemid vectors that differ only in the orientation of the cloning sites. A single-stranded version of either the sense or antisense strands of the cloned DNA fragment can be obtained, depending on the direction in which the fragment is ligated into the vector. The vectors are derived from pUC8 and pUC9 and so utilize Lac selection and carry the same polylinker as the pair of pUC vectors, enabling DNA fragments with any one of various types of sticky end to be cloned, thus providing an easy means of reversing the orientation of the cloned fragment (Section 3.2.2). Other vectors in this series (e.g. pEMBL18 and pEMBL19) have the more complex polylinkers from the higher-number pUC vectors.

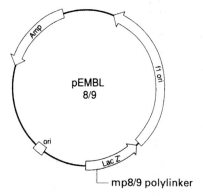

Figure 3 Structure of the phagemids pEMBL8 and pEMBL9. These vectors are 4.0 kb and carry the pMB1 and f1 origins of replication and the *amp* and *lacZ'* markers. The mp8/9 polylinkers are the same as those in pUC8/9 and M13mp8/9 (see *Figure 17*). The restriction sites that can be used for cloning are *Acc*I, *Bam*HI, *Eco*RI, *Hinc*II, *Hind*III, *Pst*I, *Sal*I, *Sma*I and *Xma*I. Reprinted from Brown, T. A., *Molecular Biology Labfax*, Vol. 1, 2nd edn, p. 306, (1998), by permission of the publisher Academic Press.

3.4 RNA expression vectors

Most plasmid vectors direct the synthesis of at least some RNA copies of the cloned DNA fragment. For example, a fragment cloned into a pUC vector is placed downstream of the *lac* promoter and is transcribed, along with the *lacZ'* gene, when the vector is introduced into the host cell. However, the *lac* promoter directs synthesis of relatively small amounts of RNA and has insufficient activity if the objective of the experiment is to obtain RNA for use as a hybridization probe (Volume II, Chapter 5) or for specialized RNA studies. For these applications, an RNA expression vector is used—one which carries a promoter sequence that is highly efficient and directs the synthesis of large amounts of RNA. The promoters that have been employed for this purpose are taken from the SP6, T3 and T7 bacteriophages. After transformation, recombinant vectors are purified, linearized by treatment with a restriction endonuclease that cuts immediately downstream of the inserted sequence (thereby defining the 3' end of the RNA that will be made) and incubated with ribonucleotides and the RNA polymerase from the appropriate phage. The phage RNA polymerase has high specificity for its promoter sequence so just one transcript is synthesized. Each polymerase can make µg amounts of RNA in this *in vitro* reaction: for SP6 RNA polymerase synthesis of 8 mol RNA/mol DNA template has been reported (18) and the T7 RNA polymerase can yield up to 30 µg of RNA/µg of DNA in 30 min (19).

3.4.1 The pSP series of RNA expression vectors

Many multipurpose vectors (Section 3.8) contain one or more bacteriophage promoters and can be used for high-level RNA synthesis. Specialized vectors that are designed primarily for this application include the pSP series marketed by Promega (*Figure 4*). Two of the vectors in the series, pSP72 and pSP73, have an

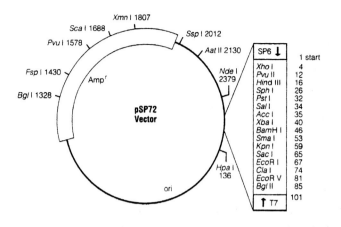

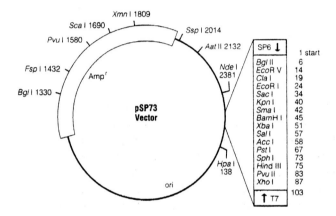

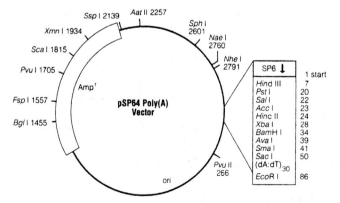

Figure 4 Structures of the pSP series of RNA expression vectors. Sizes are as follows: pSP72, 2462 bp; pSP73, 2464 bp; pSP64 Poly(A), 3033 bp. Each vector carries the pMB1 origin of replication, the *amp* marker and at least one bacteriophage promoter sequence, as indicated on the maps. The restriction sites that can be used for cloning are shown within the box on each map. Reproduced with permission from Promega Corporation.

SP6 and T7 promoter either side of a polylinker containing a number of cloning sites, the difference between the two vectors being the orientation of the poly-linker. pSP64 Poly(A) has the SP6 promoter, as well as a 30 bp dA:dT tract near the end of the polylinker. RNA synthesized from this vector therefore mimics eukaryotic mRNA in having a poly(A) tail at its 3' end.

One complication with the pSP vectors is that successful insertion of a DNA fragment into the cloning sites cannot be assayed directly. The vectors have the *amp*-selectable marker, but the polylinker is not located within this gene and ampicillin resistance is used simply to identify transformants. Recombinants can only be identified by checking the size of the purified vector DNA by agarose gel electrophoresis. This is not an oversight on the part of those who made these vectors. For many applications it is desirable that the RNA obtained has as few vector-derived sequences as possible and hence is a direct copy of just the inserted DNA. If the polylinkers in these vectors were placed within the *lacZ'* gene, for example, then recombinant selection would be more convenient but the RNA molecules obtained from the recombinant vectors would have *lacZ* sequences in their 5' regions.

3.5 Protein expression vectors

Despite the disadvantages of *E. coli* as a host for synthesis of protein from cloned genes (see *Table 1*), the ease with this bacterium can be grown, transformed and otherwise manipulated in the laboratory means that most cloning projects directed at protein synthesis utilize *E. coli*, at least in the early stages (Volume II, Chapter 8). This is particularly true if the objective of the project is not to pro-duce protein for pharmaceutical or industrial purposes, but simply to carry out structural and/or functional studies.

The synthesis of protein is more complex than the synthesis of RNA because a greater number of factors have to be taken into account when attempting to maximize the amount of product that is obtained. Rather than simply carrying out an *in vitro* experiment with a DNA molecule carrying a strong promoter—all that is required for the synthesis of large amounts of RNA—high-level protein synthesis requires that the vector carries a variety of sequence signals that are individually efficient, compatible with one another and, ideally, easily regulated. Protein synthesis occurs inside the bacterium, so steps must be taken to avoid immediate degradation of the protein by cellular activities and there must be a means of obtaining the protein in pure form from the culture. These general considerations will be covered before specific vectors for protein expression are described. The relevant issues are also discussed in Volume II, Chapter 8.

3.5.1 General features of protein expression vectors

Expression of eukaryotic genes in their natural hosts requires the coordinated action of a number of proteins, enzymes and multimolecular structures. It is mediated by a variety of nucleotide sequence motifs that lie upstream of the gene, possibly in conjunction with additional motifs downstream of the gene and others that are located some distance away (20). The entire expression system is quite different from that operating in a bacterium such as E. coli: most of the proteins and other molecules involved in expression have different structures and all the DNA motifs have different sequences. This means that if a eukaryotic gene is introduced into E. coli without modification then any protein synthesis that occurs will be entirely fortuitous, even if the gene is still linked to all the sequence motifs needed for its expression in the eukaryotic host. Protein expression vectors must therefore be designed in such a way that the eukaryotic gene, after ligation into the vector, is attached to the nucleotide sequence motifs that are required for expression in the bacterium. The most important of these motifs are:

- *The promoter* is the sequence recognized by the E. coli RNA polymerase as the position at which to bind to the DNA in order to initiate transcription of a gene. Promoters are classified as strong or weak, primarily on the basis of comparison of the amounts of gene product synthesized in the bacterium, and partly as a result of analysis of *in vitro* rates of RNA synthesis. The early protein expression vectors made use of natural E. coli promoters such as the *lac* promoter, the *trp* promoter and the p_L promoter of bacteriophage λ (which is recognized by the E. coli RNA polymerase, unlike the SP6, T3 and T7 promoters used in RNA expression vectors, each of which works with a phage-specific polymerase). During the 1980s a series of artificial promoters were synthesized by fusing components of the *lac* and *trp* sequences (e.g. the *tac* promoter; 21). An important feature of all these promoters, both natural and artificial, is that they are controllable and so expression of the cloned gene can be switched on after the transformed culture has reached a high cell density. This means that the cloned gene is expressed for a relatively short period, so breakdown of the protein product is minimized and any toxic effect that the protein has on cell metabolism is not seriously felt. The *lac* and *tac* promoters are induced by adding IPTG to the growth medium and the *trp* promoter is induced by IAA. The λp_L promoter is not chemically induced but is switched on by raising the culture temperature briefly to 42°C, which destroys a phage-encoded repressor of transcription. This repressor, the product of the *cI857* gene carried either by the vector or by the host bacterium, is mutated so that its product is temperature sensitive and inactivated by the heat shock.

- *The ribosome binding sequence* does what its name suggests and mediates attachment of the E. coli ribosome to the gene transcript. The RBS must have the appropriate sequence and be positioned some 7–9 nucleotides upstream of the initiation codon of the gene to be expressed. In some vectors a natural RBS is used, for example the one from the *lac* operon, and in other vectors a synthetic sequence is incorporated at the relevant position.

If the gene to be expressed is ligated at the correct position downstream of a promoter and RBS that are active in E. coli, protein synthesis will occur. If the promoter is switched on at the appropriate point of the growth cycle then there will be only limited opportunity for the protein to be degraded before the cells are harvested, and degradation can be further reduced by using an E. coli strain that carries a mutation such as *lon* which inactivates one or more of the main cellular proteases. The final consideration is how to purify the protein from the culture. This could be achieved by making a protein extract from the bacteria, and then using conventional methods to separate the cloned protein from the natural E. coli ones. This is feasible because the cloned protein often makes up over 40% of the total protein content of the cell, but the procedure is time-consuming and tedious. For this reason, most modern protein synthesis vectors are designed in such a way that the protein is expressed as a fusion with a second protein, the latter offering some means for convenient purification from the protein extract (usually by an affinity-binding procedure). This approach has the added advantage that if the second protein is a natural E. coli one then the fusion protein will be more stable than the unfused protein because the bacteria will not recognize it as being foreign. Examples of fusion systems used in protein expression are given in the descriptions of specific vectors that follow.

3.5.2 Examples of protein expression vectors

The range of protein expression vectors available to the molecular biologist is vast, with over 30 types in common usage (1). Three will be described to illustrate the various options.

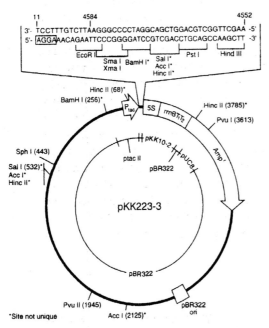

Figure 5. Structure of the protein expression vector pKK223-3. The vector is 4584 bp and carries the pMB1 origin of replication and the *amp* marker. The restriction sites that can be used for cloning are *Eco*RI, *Hind*III, *Pst*I, *Sma*I and *Xma*I. These sites are located downstream of the *tac* promoter and upstream of the termination sequences from an *E. coli* ribosomal RNA gene. The RBS is shown in a box in the sequence given at the top of the map. The thin circle inside the main map shows the sources of the segments that were linked together to construct this vector. Reproduced with the kind permission of Amersham Pharmacia Biotech AB.

The first of these is pKK223-3 (*Figure 5*; available from Pharmacia), which was one of the earliest protein expression vectors to be constructed (22). It carries a polylinker that contains an RBS located upstream of a variety of cloning sites. Transcription is initiated from the *tac* promoter and terminates at signals taken from one of the *E. coli* ribosomal RNA operons. The vector can be used to obtain high levels of protein synthesis, but there is no fusion system and proteins must purified by conventional means from cell extracts.

The pGEX series of vectors (*Figure 6*; available from Pharmacia) are among the most popular expression vectors that give a fusion product (23). The *tac* promoter is again used, but this time it is placed upstream of a lengthy open reading frame coding for most of the GST gene of *E. coli*. The cloning sites lie immediately downstream of this open reading frame, arranged in such a way that insertion of an appropriate DNA fragment leads to synthesis of a fusion protein in which the GST and cloned protein components are separated by a short amino acid sequence that is a recognition site for either the thrombin or Factor Xa pro-teases. Purification of the fusion protein makes use of the high affinity of GST for its natural substrate, glutathione (for example, by passing the protein extract through a chromatography column packed with glutathione-agarose). The fusion protein is then cleaved by treatment with the protease, enabling the cloned protein to be recovered free from its GST partner (see Volume II, Chapter 8, *Protocols 2* and *3*).

The pET vectors simplify protein expression even further (24). With some of these vectors the fusion is with a short peptide containing six histidine residues. This tag enables the fusion protein to be purified by affinity chromatography with a resin containing NTA, which chelates a metal ion that forms a bridge between the resin and the histidine tag (25). The tag is so short that in many cases it is not necessary to cleave it off of the cloned protein. One feature to note about these vectors is that they do not use an *E. coli* promoter and instead contain the promoter from gene 10 of bacteriophage T7. They therefore have to be used with a special host strain that carries the gene for the T7 RNA polymer-ase (see *Appendix 1*). Regulation of protein synthesis is possible because this gene is placed downstream of the *tac* promoter, so induction with IPTG switches on expression of the T7 RNA polymerase which, in turn, activates transcription of the gene inserted into the vector.

3.6 Cosmids

Cosmids combine features of plasmids and bacteriophage λ vectors and are used to clone large pieces of DNA, up to 45 kb long, when genomic libraries are prepared (Volume II, Chapter 2). Their plasmid features enable the vectors to replicate at high copy number and to be amplified in the presence of chlor-amphenicol, and provide a means for selection of transformed and recombinant cells. Since they also contain λ *cos* sequences, they can undergo *in vitro* pack-aging for efficient recovery of recombinant vectors (see Chapter 8, Section 4.1). After ligation of the vector to the DNA fragments to be cloned, concatamers of

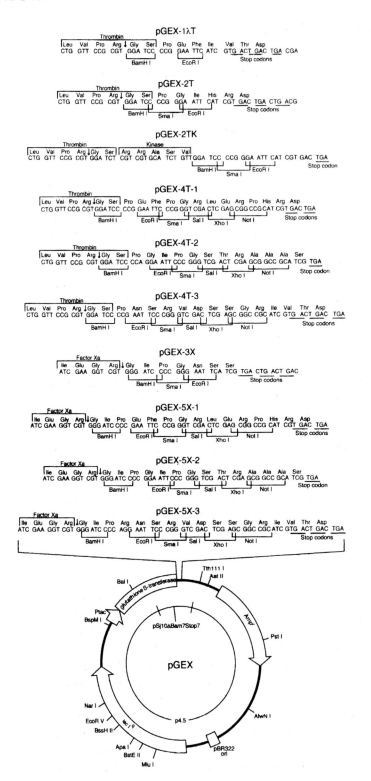

Figure 6 Structures of the pGEX series of protein expression vectors. These vectors are approximately 4950 bp and each carries the pMB1 origin of replication, the *amp* marker and the *tac* promoter. The restriction sites that can be used for cloning with the different vectors in the series are shown above the circular plasmid map. The appropriate constructions lead to synthesis of a GST fusion protein, which can be cleaved by treatment with either thrombin or Factor Xa, as indicated on the maps. The vector also carries *lacI^q*, a mutated version of the *lac* repressor gene. This mutation results in overexpression of the *lac* repressor and reduces the background level of expression of the fusion protein in uninduced cells. The thin circle inside the main map shows the sources of the segments that were linked together to construct this vector. Reproduced with the kind permission of Amersham Pharmacia Biotech AB.

vector and DNA inserts are formed; these act as substrates for *in vitro* packaging. The resulting phage particles can be used to infect *E. coli* cells, within which the recombinant cosmids are propagated as plasmids.

Most cosmids are relatively small (≈ 5 kb), and so can accommodate up to 45 kb of exogenous DNA before reaching the upper DNA-carrying capacity of the λ phage particle. This is a significantly greater length of DNA than can be cloned in a conventional λ vector (Section 4). The larger size of the inserts reduces the number of cosmids needed for a representative gene library and enhances the likelihood that a cloned gene will be present as an intact copy in a cosmid library if this gene is relatively long, as is the case for most eukaryotic genes.

The advantage provided by the high capacity of cosmids is slightly offset by a few disadvantages displayed by these vectors. The efficiency of the *in vitro* packaging system is 100- to 1000-fold lower for cosmid vectors than for phage vectors, and the colony hybridization methods (Volume II, Chapter 5) used to screen clone libraries for a recombinant containing the desired gene are less effective with a cosmid library. This is largely due to the much lower concentrations of recombinant cosmids in an *E. coli* colony as a result of the reduced copy number of these very large plasmids, compared with the concentration of recombinant phage in a plaque. Finally, preparation of a cosmid library requires starting DNA that is much larger (> 100 kb) than is necessary for construction of a phage library.

3.6.1 Examples of cosmid vectors

Plasmid pJB8 (*Figure 7*) (26), which has been extensively used as a cosmid vector, is based on a pMB1 replicon, carries one *cos* sequence and uses the *amp* gene as a selectable marker. The vector is generally used for the cloning of size-selected *Sau*3AI-or *Mbo*I-digested genomic DNA at the unique *Bam*HI site. After *in vitro* packaging and infection of *E. coli*, recombinant cosmid DNA can be amplified with chloramphenicol (*Protocol 1*), to improve the efficiency of library screening by colony hybridization.

The pcosEMBL vectors (*Figure 8*) (27) use the R6K replicon and the *kan* and *tet* selectable markers; the latter is inactivated by insertion of new DNA into the single *Bam*HI site. The vectors are deliberately designed to have components as different as possible from those possessed by the pUC vectors. There is no

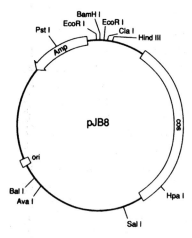

Figure 7 Structure of the cosmid pJB8. The vector is 5.4 kb and carries the pMB1 origin of replication, the *amp* marker and a λ *cos* sequence. The restriction sites that can be used for cloning are *Bam*HI, *Cla*I, *Eco*RI, *Hin*dIII and *Sal*I. Reprinted from Brown, T. A., *Molecular Biology Labfax*, Vol. 1, 2nd edn, p. 317, (1998), by permission of the publisher Academic Press.

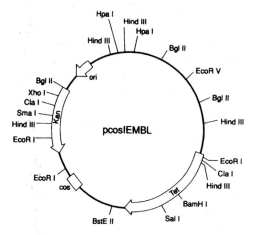

Figure 8 Structure of the cosmid pcos1EMBL. The vector is 6.1 kb and carries the R6K origin of replication, the *kan* and *tet* markers and a λ *cos* sequence. The restriction sites that can be used for cloning are *Bam*HI, *Sal*I, *Sma*I and *Xho*I. pcos2EMBL has a similar structure but has two *cos* sequences, which simplifies preparation of the cosmid prior to ligation with new DNA. Reprinted from Brown, T. A., *Molecular Biology Labfax*, Vol. 1, 2nd edn, p. 304, (1998), by permission of the publisher Academic Press.

significant nucleotide sequence identity between pcosEMBL and pUC, which means that a pUC recombinant can be used to screen a pcosEMBL library without the result being swamped by hybridization between vector sequences. So, for example, a pUC vector carrying a cDNA can be used directly to screen a pcoEMBL library for a cosmid clone containing the equivalent gene.

3.7 Cloning vectors for PCR products

The central position taken by PCR in modern molecular biology has led to the development of many support technologies for the further analysis of PCR products (Volume II, Chapter 7). Among these support technologies are the special class of plasmid vectors designed specifically for the cloning of PCR products. The double-stranded DNA products resulting from amplification with *Taq* DNA polymerase are unusual in that they are not perfectly blunt-ended. Instead, a proportion of the ends have a single adenosine extension at the 3'-terminus, which prevents the immediate ligation of the PCR product into a blunt-ended cloning vector. It is possible to convert the A-overhang into a genuine blunt end by treatment with exonuclease I but this procedure increases the time that must be spent on the project and, in any case, is not entirely efficient. An alternative would be to use a cloning vector that possesses a restriction site which, after cleavage, gives sticky ends consisting of single 3'-thymidines. This is a possibility but is not the standard method used simply because there are no commonly available restriction endonucleases that leave such an sticky end. Instead, the usual solution is to use a plasmid vector that is has been linearized with a blunt-end restriction enzyme and then further treated to add T-overhangs to the free 3'-ends that are created. This is in itself is a difficult manipulation but commercial suppliers of cloning vectors are very willing to devote the necessary time and expense to the task (see Chapter 7, Section 4.4).

pGEM-T (*Figure 9*) and pGEM-T Easy (both marketed by Promega) are examples of this type of cloning vector. The two vectors differ only in the identities of the restriction sites that flank the cloning site and which can be used to excise the PCR product as a sticky-ended fragment after it has been ligated into the pGEM

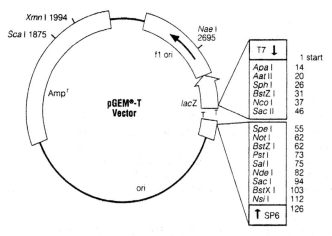

Figure 9 Structure of the pGEM-T vector for cloning pCR products. The vector is 3003 bp and carries the pMB1 and f1 origins of replication, the *amp* and *lacZ'* markers and the SP6 and T7 promoter sequences. The vector is supplied in linear form with single thymidine overhangs at the 3' ends. After cloning, the restriction sites shown in the boxes can be used to re-excise the PCR product. Reproduced with permission from Promega Corporation.

vector and propagated in *E. coli*. In fact, excision of the cloned fragment and reinsertion into a second vector is rarely necessary because pGEM-T and pGEM-T Easy are multipurpose vectors that also allow the synthesis of a single-stranded DNA version of the cloned PCR product and the expression of RNA from T7 and SP6 promoter sequences that lie either side of the insertion site. Other vectors for cloning PCR products include the pCR series (Stratagene) and the pDIRECT series (Clontech).

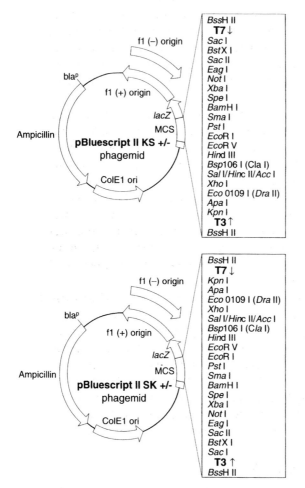

Figure 10 Structure of the multipurpose cloning vector pBluescript II. The vector is 2961 bp and carries the ColE1 and f1 origins of replication, the *amp* and *lacZ'* markers and the T3 and T7 promoter sequences. The restriction sites in the multi-cloning sequence (MCS) are shown in the boxes. The vector comes in four forms. The SK or KS versions differ in the orientations of the multiple cloning site, and the + and − versions have the f1 origin in opposite directions, enabling either the + or − polynucleotide of the plasmid to be obtained as single-stranded DNA. The label 'bla^p' indicates the promoter of the *amp* gene. pBluescript SK+/− is an older vector pair with similar properties to pBluescript II. Reproduced with permission from Stratagene.

3.8 Multipurpose vectors

During the 1970s and 1980s the majority of cloning experiments were carried out with vectors that had been passed between molecular biology research groups on a freeshare basis. During the 1990s this spirit of scientific partnership was largely overtaken by the market economy, with most molecular biologists obtaining their vectors from commercial suppliers. In fact, this is a much better system because it means that the quality of your vectors is assured and the vectors themselves are more sophisticated. The trend among commercial suppliers has been to develop multipurpose vectors that can be used for any of several tasks and which fall into several of the categories considered so far in this section. Individual vectors that combine the properties of phagemids and RNA expression vectors, and which are equally effective as basic cloning vectors, are available from several different companies; a few vectors are phagemids that can be used for both RNA and protein expression.

There are many different examples of multipurpose vectors available from different companies. Two series will be described here. The first is pBluescript (28), marketed by Stratagene, which has an impressive battery of restriction sites, enabling fragments with most possible sticky ends to be ligated into the *lacZ'* gene (*Figure 10*). The vectors have an f1 origin of replication and so can generate single-stranded DNA, and the cloning sites are flanked by T3 and T7 promoters for RNA expression. The vectors in the series differ in the identities of the cloning sites, in the order of these sites relative to the two bacteriophage promoters and in the orientation of the f1 origin of replication (the last property means that either strand of the double-stranded plasmid can be recovered as single-stranded DNA).

The pGEM vectors (Promega) are equally versatile. Two of these, pGEM-T and pGEM-T Easy, have already been mentioned (Section 3.7) as multipurpose vectors with the special ability of being able to clone PCR products. Two other vectors in this series, pGEM-3Z and pGEM-4Z, are straightforward RNA expression vectors. The other 10 vectors make up five pairs whose properties are illustrated by pGEM-3Zf(+/−) (*Figure 11*). As with the pBluescript vectors, the pGEM series offer the possibility of cloning fragments with many different types of sticky end, so that the inserted DNA is placed under the control of the SP6 and T7 promoters and can be obtained as single-stranded DNA of either the + or − polarity.

4 Bacteriophage λ vectors

E. coli viral vectors containing double-stranded DNA are all derived from the temperate bacteriophage called λ. Wild-type λ DNA contains several target sites for most commonly-used restriction enzymes, but derivatives have been produced (by *in vitro* mutagenesis and other techniques) that lack these multiple restriction sites and so can be cut at specific positions prior to the addition of new DNA. The λ genome is 48.5 kb and the λ bacteriophage particle can accommodate up to 52 kb of DNA so, at first, it might appear difficult to clone anything other than very short fragments in a λ vector. However, various parts of

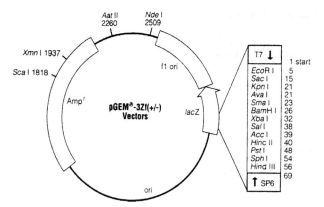

Figure 11 Structure of the multipurpose cloning vector pGEM-3Zf(+/−). The vector is 3199 bp and carries the pMB1 and f1 origins of replication, the *amp* and *lacZ'* markers and the SP6 and T7 promoter sequences. The restriction sites that can be used for cloning are shown in the box. The + and − versions of the vector have the f1 origin in opposite orientations, enabling either the + or − polynucleotide of the plasmid to be obtained as single-stranded DNA. Reproduced with permission from Promega Corporation.

the λ genome are required solely for the lysogenic life cycle (Chapter 2, Section 5.1), which is unnecessary when the bacteriophage is used as a cloning vector. Deletion of one or more of these dispensable regions decreases the length of the cloning vector and increases the size of the DNA that can be inserted before the upper limit for packaging in the phage particle is reached. Unfortunately, there is also a lower limit for the size of the molecule that can successfully be packaged ($\approx$ 38 kb); the figure is relevant because the non-recombinant vector must be longer if it is to be propagated prior to use in cloning. These considerations have led to development of two types of λ vector:

- Insertion vectors are ones which simply have a unique restriction site into which new DNA is inserted. The non-recombinant vector must be > 38 kb, so the largest fragment that can be cloned in a vector of this type is about 14 kb.

- Replacement vectors contain a pair of restriction sites flanking a dispensable region (or 'stuffer fragment') that is replaced by the DNA to be cloned. The presence of the stuffer fragment ensures that the non-recombinant vector is > 38 kb, even though extensive regions of the λ genome have been deleted. If all possible deletions are made then the vector sequences are about 28 kb and up to 24 kb of new DNA can be cloned.

The cloning capacity of λ vectors therefore falls between that of plasmid vectors, which are unstable if fragments of > 10 kb are inserted, and cosmids, which are used to clone DNA in the 35–45 kb range.

4.1 Practical considerations relevant to the use of λ vectors

The strategy for the use of a replacement vector to prepare a gene library is illustrated in *Figure 12*. By treatment with an appropriate restriction enzyme,

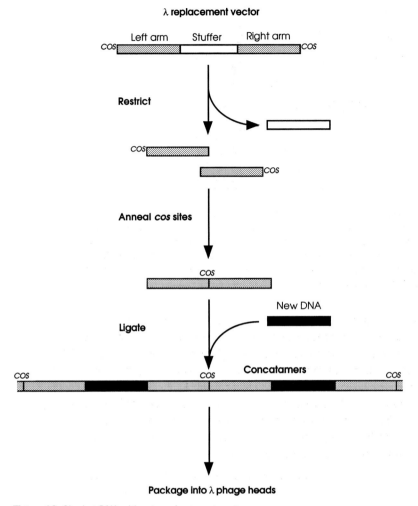

Figure 12 Cloning DNA with a λ replacement vector.

the vector DNA is split into three parts. One part, called the left arm, contains all the functions for morphogenesis of the virus particle, while a second fragment, called the right arm, harbours all functions for multiplication of viral DNA and expression of viral genes. The central stuffer fragment contains no essential functions and is dispensable. In the commonly-used replacement vectors, nucleotide sequences coding for one or more restriction enzyme sites have been introduced at both sides of the stuffer fragment, so this fragment can simply be uncoupled from the vector arms by treatment with a single restriction enzyme. Phage λ DNA contains 5′-protruding ends of 12 nucleotides, each of which can anneal to one another, forming a duplex molecule in which the left and right arms are joined by these cohesive ends (*cos* sites). When an exogenous DNA fragment is ligated to the vector, a concatamer of vector and exogenous DNA is generated. Viable phage can then be formed from these concatamers in an *in vitro* packaging system. In order to be packaged into viable phage, the DNA segment between the two *cos* sites should be larger than 38 kb but smaller than 52 kb. These are the lower and upper limits, respectively, for the size of a λ genome, in order to be encapsulated into a phage head. The vector fragment comprising left and right arms without the stuffer fragment is much smaller than 38 kb. Consequently, it will not be packaged into phage heads unless an exogenous DNA fragment has been ligated to it. Detailed protocols for the use of a λ replacement vector in construction of a genomic library are given in Volume II, Chapter 2.

Insertion vectors are used for the introduction of small DNA fragments, ranging in size between 1 and 10 kb. An important difference between insertion vectors and replacement vectors is that the two fragments, after cleavage of an insertion vector with a restriction enzyme, can form viable phages when no exogenous DNA has been incorporated, whereas a replacement vector from which the stuffer fragment has been removed cannot form viable phages. Thus, there is a natural selection against parental replacement vectors, but special selection systems have to be introduced to discriminate between recombinant and parental versions of an insertion vector. The special selection is usually based on insertional inactivation of a phage gene or a gene that was purposely introduced into the vector, leading to a change in phenotype (see Chapter 8, Section 3.4). If, for example, exogenous DNA is cloned into a site within the immunity region, as is the case with the 'immunity insertion vectors' (29), recombinants can be easily and efficiently selected. First, recombinant phage multiply faster than parent phage because of their enlarged genome. They form larger plaques and rapidly outgrow the parent phage. More importantly, recombinant phage can be identified by eye because inactivation of the immunity region results in clear plaques, while the parent phage shows the characteristic turbid plaque morphology resulting from growth in the plaque area of bacteria that have undergone the lysogenic response to phage infection. Lysogenic bacteria produce the λ repressor (product of gene *c*I) that negatively controls the expression of other phage genes, thereby preventing the lytic cycle of the phage. In *E. coli* carrying the high frequency of lysogeny mutation (Hfl⁻), the λ repressor is overproduced to the extent that plaque formation is suppressed (30). However, *c*I⁻ phage forms plaques with normal efficiency on an Hfl⁻ strain, Recombinant phage made from immunity insertion vectors can thus readily be selected on such a strain.

Recombinant phage can also be selected on the basis of their ability to multiply in bacteria lysogenic for phage P2 (31). Such phage are said to have the Spi⁻ phenotype (lack of sensitivity to P2 interference) (32). Wild-type phage λ cannot grow on such a host, but mutants from which the *red* and *gam* genes in the non-essential region of the genome have been removed or inactivated and have incorporated an octanucleotide sequence called Chi, normally produce plaques on a P2 lysogen. If the stuffer fragment of a replacement vector carries the Chi sequence as well as functional *red* and *gam* genes, then it is not necessary to remove the stuffer fragment from the restriction digest before ligating the vector arms to the exogenous DNA fragment. If the resulting phage are plated on a P2 lysogenic host, only recombinants that are Red⁻Gam⁻Chi⁺ will produce plaques, while the parent phages will not because they are Red⁺Gam⁺.

4.2 Examples of λ vectors

4.2.1 λgt vectors

The λgt series (*Figure 13*) are insertion vectors specifically designed for the cloning of relatively short DNA fragments, in particular cDNAs (see Volume II, Chapter 3). λgt10 can accept DNA fragments of up to 6 kb into the single *Eco*RI cloning site (33). Insertion inactivates the λ repressor gene, generating a *cI*⁻ phage, which forms a clear plaque. Non-recombinant λgt10 forms turbid plaques. Since λgt10 does not require an insert fragment to yield a packageable DNA molecule, cDNA libraries cloned in λgt10 initially consist predominantly of parent phage if plated on wild-type *E. coli* host bacteria. Selection against the parent phage is possible during amplification of the library by plating the phage on a Hfl⁻ strain, which permits plaque formation of *cI*⁻ phage only.

A second member of this vector series, λgt11, is also very useful for the construction of cDNA libraries (34). The vector was designed to allow insertion of DNA fragments up to 7.2 kb in a unique *Eco*RI site located near the C-terminal end of the *E. coli lacZ* gene. A DNA segment which is properly aligned with the reading frame of the *lacZ* gene will be expressed as a fusion protein. A characteristic feature of the vector is its ability to express a cloned DNA fragment, enabling the recombinant vector to be distinguished from the parent one on the basis of the product that is formed. The presence of a gene in a λgt11 clone library can easily be detected by binding of the protein expressed from the cloned gene to antibodies raised against it. To maximize genetic stability and phage yield, recombinant phage are propagated on the special host strain, *E. coli* Y1088 (see *Appendix 1*). The *lac* deletion in this strain prevents host–phage recombination and *supF* suppresses the *S*100 amber mutation carried by the phage. To minimize degradation of the fusion proteins that are formed, a protease-minus (*lon*) strain, Y1090 (*Appendix 1*), can be used for expression studies.

A third vector, λgt18 (35), harbours a polylinker region with unique cloning sites for *Eco*RI and *Sal*I, making the vector somewhat more versatile for cloning cDNAs. To further improve the usefulness of the vector, a Chi sequence was inserted into the vector as well as a unique *Not*I site in the polylinker region, yielding λgt22 (36).

4.2.2 The λEMBL vectors

The λEMBL phage (37) are a family of replacement vectors with the generally useful properties of high capacity, polylinker cloning sites and genetic selection for recombinant phage. λEMBL3 and λEMBL4 (*Figure 14*) are particularly useful for the construction of gene banks by ligating partial *Sau*3AI digests of genomic DNA into the *Bam*HI sites of the vectors (see Volume II, Chapter 2). Alternatively,

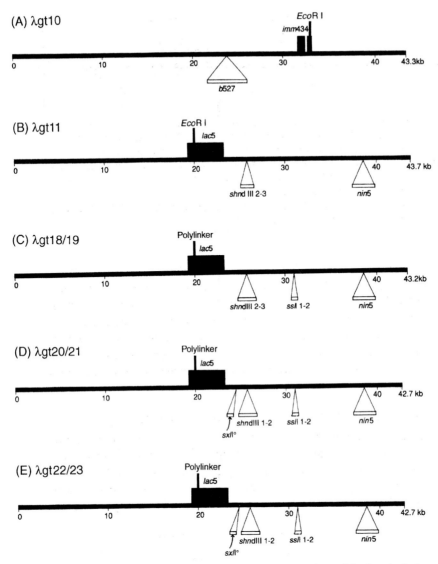

Figure 13 Structures of the λgt insertion vectors. The maximum sizes of the inserts that can be cloned are: λgt10, 6.0 kb; λgt11, 7.2 kb; λgt18 and λgt19, 7.7 kb; λgt20 and λgt21, 8.2 kb; λgt22 and λgt23, 8.2 kb. With λgt10, recombinants are identified on the basis of their *cl* phenotype; with the other vectors Lac selection is used. The restriction sites used for cloning are: (A) λgt10, *Eco*RI; (B) λgt11, *Eco*RI; (C) λgt18 and λgt19, *Eco*RI and *Sal*I; (D) λgt20 and λgt21, *Eco*RI, *Sal*I and *Xba*I; (E) λgt22 and λgt23, *Eco*RI, *Not*I, *Sac*I, *Sal*I and *Xba*I. The members of the vector pairs (λgt18/19, 20/21 and 22/23) differ in the relative positions of the restriction sites in the polylinkers. An additional vector not shown here, λgt22A, is the same as λgt22 but with a *Spe*I site instead of the *Sac*I site. Solid boxes above the map indicate DNA sequences from other sources that were inserted into these vectors during their construction and the open boxes below the map are deletions of the wild-type λ genome (see ref. 1 for a full description). Reprinted from Brown, T. A., *Molecular Biology Labfax*, Vol. 1, 2nd edn, pp. 354–358, (1998), by permission of the publisher Academic Press.

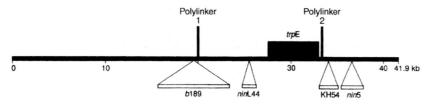

Figure 14 Structures of the λEMBL3 and λEMBL4 replacement vectors. Inserts of 7–20 kb can be cloned in the *Bam*HI, *Eco*RI or *Sal*I restriction sites. Recombinants are identified by virtue of their Spi⁻ phenotype. The two vectors differ in the relative positions of the restriction sites in the polylinkers. Solid boxes above the map indicate DNA sequences from other sources that were inserted into these vectors during their construction and the open boxes below the map are deletions of the wild-type λ genome (see ref. 1 for a full description). Reprinted from Brown, T. A., *Molecular Biology Labfax*, Vol. 1, 2nd edn, p. 351, (1998), by permission of the publisher Academic Press.

DNA fragments with cohesive ends matching those of *Bam*HI, *Eco*RI or *Sal*I can be inserted into the vector. The size of the fragments that can be incorporated is minimally 7 kb and maximally 20 kb. When the vector is treated with either *Bam*HI, *Eco*RI or *Sal*I, the central fragment can easily be removed and, because of their Spi⁻ phenotype, recombinant phage can be selected on P2 lysogenic host bacteria. In λEMBL3 the *Sal*I sites in the polylinker region are nearest to the vector arms, whereas λEMBL4 has the linkers reversed and the *Eco*RI sites outermost.

4.2.3 Commercial multipurpose λ replacement vectors

As with plasmid vectors (Section 3.8), the trend in recent years has been towards commercial λ vectors that provide diverse functions. A pair of vectors marketed by Stratagene, λDASHII and λFIXII (*Figure 15*), are replacement vectors with a wide range of cloning sites and a capacity for inserted DNA fragments of 9–22 kb. Recombinants are identified by virtue of their Spi⁻ phenotype. The two

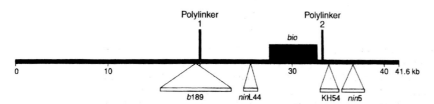

Figure 15 Structure of the multipurpose λ replacement vector λDASHII. The vector can be used to clone fragments of 9–22 kb in the *Bam*HI, *Eco*RI, *Hind*III, *Not*I, *Sac*I, *Sal*I, *Xba*I or *Xho*I restriction sites. Polylinker 1 contains (in addition to the restriction sites) the T3 promoter sequence and polylinker 2 contains the T7 promoter sequence. Recombinants are identified by virtue of their Spi⁻ phenotype. λFIXII has an identical structure to λDASHII, except that the positions of the promoter sequences are interchanged. Solid boxes above the map indicate DNA sequences from other sources that were inserted into these vectors during their construction and the open boxes below the map are deletions of the wild-type λ genome (see ref. 1 for a full description). Reproduced with permission from Stratagene.

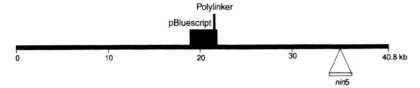

Figure 16 Structure of the multipurpose λ insertion vector λZAPII. The vector can be used to clone fragments of up to 10 kb in the *Eco*RI, *Not*I, *Sac*I, *Spe*I, *Xba*I or *Xho*I restriction sites. The polylinker is flanked by the T3 and T7 promoter sequences and lies within a pBluescript plasmid sequence, which means that recombinants are identified by Lac selection and cloned DNA can be recovered in single-stranded form. Solid boxes above the map indicate DNA sequences from other sources that were inserted into this vector during its construction and the open boxes below the map are deletions of the wild-type λ genome (see ref. 1 for a full description). Reproduced with permission from Stratagene.

vectors have slightly different polylinker sequences but both also include T7 and SP6 promoter sequences and so can be used to obtain RNA transcripts of the inserted DNA. A third Stratagene product, λZAPII (*Figure 16*), is an insertion vector with a maximum capacity of 10 kb with recombinants identified by Lac selection. This vector not only has a range of cloning sites and the pair of bacteriophage promoters, it also harbours a complete copy of the plasmid vector pBluescript SK– (see *Figure 10*), into which new DNA is inserted, once again enabling recombinants to be identified by Lac selection. The pBluescript sequence is flanked on one side by the initiator region and on the other by the terminator region of the origin of replication of bacteriophage f1, which means that the plasmid and insert can be recovered as single-stranded DNA.

5 Bacteriophage M13 vectors

The need to obtain single-stranded versions of cloned DNA fragments for DNA sequencing was initially satisfied by the development of vectors based on bacteriophage M13. Although these vectors have a relatively low upper limit for cloned DNA fragments they are still frequently used to produce template DNA for sequencing experiments (Volume II, Chapter 6). Their popularity has mainly been due to the simplicity and efficiency of the methods for preparing single-stranded DNA from the secreted phage particles: a 5 ml culture yields sufficient high-quality DNA for several sequencing runs. Even now, when the use of M13 in large-scale sequencing has largely been superseded by the advent of thermal cycling methods, these vectors are still routinely employed in small-scale projects and for other applications that require single-stranded DNA.

5.1 General properties of M13 vectors

The genomes of filamentous phages such as M13, f1 and fd consist of circular single-stranded DNA. Upon infection of sensitive, F⁺ *E. coli* bacteria, the phage

DNA is converted into a double-stranded replicative form (RF), which is extensively replicated to yield a pool of double-stranded phage DNA. At the end of the replicative cycle, the formation of single-stranded viral DNA ensues which eventually results in the formation of progeny phages; however, this is not a lytic infection process, as the bacteria continue to multiply for a number of generations while extruding phage into the culture fluid. Because the concentrations of RF DNA and of progeny phage produced are very high (10^{11}–10^{12} phage particles/ml), it is relatively easy to isolate large quantities of both double- and single-stranded versions of the phage genome. New DNA that is inserted into a non-essential region of the double-stranded genome, using the standard

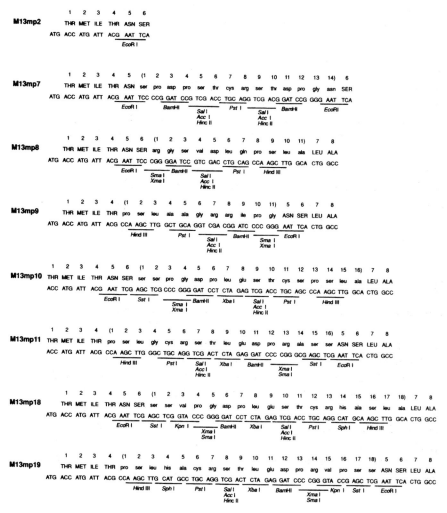

Figure 17 The polylinker sequences of the M13mp series of single-stranded bacteriophage vectors. Each vector is approximately 7.2 kb, with the polylinker sequence located within the *lacZ'* gene. Reprinted from Brown, T. A., *Molecular Biology Labfax*, Vol. 1, 2nd edn, p. 363, (1998), by permission of the publisher Academic Press.

approach based on restriction and ligation of circular vector DNA (Chapter 7), is replicated along with rest of the genome after transformation of host *E. coli* bacteria. The filamentous phage capsid is little more than a coating of proteins and, in contrast to the situation with bacteriophage λ, does not present a barrier to the accommodation of extra DNA, although the upper limit lies at about 3 kb of inserted DNA because larger DNA inserts are unstable (38).

Of the three filamentous phages mentioned above, M13 is used most frequently for the construction of single-stranded recombinant DNA phages. The second phage, fl, has proven more useful as the donor of the origin of single-stranded DNA replication in phagemid vectors (Section 3.3). Vectors derived from the third type of phage, fd, have also been used in the past but are now almost unheard of.

5.1.1 The M13mp series

The vast majority of cloning experiments with a single-stranded bacteriophage use one of the M13mp series of vectors (38–40). Each of these vectors carries a copy of the *lacZ'* gene located in the so-called intergenic region of the phage genome. The vectors differ in the polylinker cloning sequences that have been placed in the *lacZ'* gene, which gradually become more complex as the number of the vector increases (*Figure 17*). In M13mp7 the polylinker is flanked by a pair of *Eco*RI sites, but in the vector pairs M13mp8/9, M13mp10/11 and M13mp18/19 one end of the polylinker is marked by an *Eco*RI site and the other by a *Hind*III sequence. Inserts can therefore be 'turned around' by excising from one member of a vector pair by *Eco*RI–*Hind*III restriction followed by ligation into the other member of the pair. This results in the orientation of the insert in the vector being reversed and allows both ends of a cloned fragment to be sequenced using the same primer (see Volume II, Chapter 6). M13mp10 and M13mp11 carry amber mutations in essential genes which facilitates selection of recombinant phage after *in vitro* mutagenesis of an inserted DNA fragment (39).

Acknowledgements

Some parts of this chapter, notably Sections 2, 3.1, 4 and 5, are based on the equivalent chapter from *Essential Molecular Biology: a Practical Approach,* 1st edn, by P. H. Pouwels, published by Elsevier Science Ltd, Amsterdam.

References

1. Brown, T. A. (ed.) (1998). *Molecular biology labfax*, 2nd edn, Vol. 1. Academic Press, London.
2. Pouwels, P. H., Enger-Valk, B. E., and Brammar, W. J. (1985). *Cloning vectors*. Elsevier Science Publishers, Amsterdam.
3. Warren, G. J., Twigg, A. J., and Sheratt, D. J. (1978). *Nature* **274**, 259.
4. Meacock, P. A. and Cohen, S. N. (1980). *Cell* **20**, 529.
5. Summers, D. K. and Sherratt, D. J. (1984). *Cell* **36**, 1097.
6. Cesareni, G., Muesing, M. A., and Poliski, B. (1982). *Proc. Natl. Acad. Sci. USA* **79**, 6313.

7. Beebee, T. J. C. and Burke, J. (1992). *Gene structure and transcription: in focus*. BIOS, Oxford.

8. Ptashne, M. (1992). *A genetic switch: phage λ and higher organisms*, 2nd edn. Blackwell, Oxford.

9. Bolivar, F., Rodriguez, R. L., Greene, P. J., Betlach, M. C., Heyneker, H. L., and Boyer, H. W. (1977). *Gene* **2**, 95.

10. Summers, D. K. and Sherratt, D. J. (1984). *Cell* **36**, 1097.

11. Soberon, X., Covarrubias, L., and Bolivar, F. (1980). *Gene* **9**, 287.

12. Zurita, M., Bolivar, F., and Soberon, X. (1984). *Gene* **28**, 119.

13. Vieira, J. and Messing, J. (1982). *Gene* **19**, 259.

14. Norrander, J., Kempe, T., and Messing, J. (1983). *Gene* **26**, 101.

15. Dotto, G. P. and Zinder, N. D. (1984). *Nature* **311**, 279.

16. Dotto, G. P., Enea, V., and Zinder, N. D. (1981). *Virology* **114**, 463.

17. Dente, L., Cesarini, G., and Cortese, R. (1983). *Nucl. Acids Res.* **11**, 1645.

18. Johnson, M. T. and Johnson, B. (1984). *Biotechniques* **2**, 156.

19. Dunn, J. J. and Studier, F. W. (1983). *J. Mol. Biol.* **166**, 477.

20. Brown, T. A. (1999). *Genomes*. BIOS, Oxford.

21. Amann, E., Brosius, J., and Ptashne, M. (1983). *Gene* **25**, 167.

22. Brosius, J. (1984). *Gene* **27**, 151.

23. Smith, D. B. and Johnson, K. S. (1988). *Gene* **67**, 31.

24. Studier, F. W. and Moffat, B. A. (1986). *J. Mol. Biol.* **189**, 113.

25. Hochuli, E., Bannwarth, W. Döbeli, H., Gentz, R., and Stüber, D. (1988). *Biotechnology* **6**, 1321.

26. Ish-Horowicz, D. and Burke, J. F. (1981). *Nucl. Acids Res.* **9**, 2989.

27. Poustka, A., Rackwitz, H.-R., Frischauf, A. M., Hohn, B., and Lehrach, H. (1984). *Proc. Natl. Acad. Sci. USA* **81**, 4129.

28. Short, J. M., Fernandez, J. M., Sorge, J. A., and Huse, W. D. (1988). *Nucl. Acids Res.* **16**, 7583.

29. Murray, N. E., Brammar, W. J., and Murray, K. (1977). *Mol. Gen. Genet.* **150**, 53.

30. Hoyt, M. A., Knight, D. M., Das, A., Miller, H. I., and Echols, H. (1982). *Cell* **31**, 565.

31. Hendrix, R. W., Roberts, J. W., Stahl, F., and Weisberg, R. A. (ed.) (1983). *Lambda II*. Cold Spring Harbor Laboratory Press, Cold Spring Harbor.

32. Zissler, J., Signer, E., and Shaefer, F. (1971). In *The bacteriophage lambda* (ed. A. D. Hershey), p. 469. Cold Spring Harbor Laboratory Press, Cold Spring Harbor.

33. Huynh, T. V., Young, R. A., and Davis, R. W. (1985). In *DNA cloning techniques: a practical approach* (ed. D. M. Glover), p. 49. IRL Press, Oxford.

34. Young, R. A. and Davis, R. W. (1983). *Proc. Natl. Acad. Sci. USA* **80**, 1194.

35. Han, J. H., Stratowa, C., and Rutter, W. J. (1987). *Biochemistry*, **26**, 1617.

36. Han, J. H. and Rutter, W. J. (1987). *Nucl. Acids Res.*, **15**, 6304.

37. Frischauff, A. M., Lehrach, H., Poustka, A., and Murray, N. (1983). *J. Mol. Biol.* **170**, 827.

38. Yannisch-Perron, C., Vieira, J., and Messing, J. (1985). *Gene* **33**, 103.

39. Messing, J. (1983). In *Methods in enzymology* (ed. R. Wu, L. Grossman, and K. Moldave), Vol. 101, p. 20. Academic Press, London.

40. Messing, J., Gronenborn, B., Muller-Hill, B., and Hofschneider, P. H. (1977). *Proc. Natl. Acad. Sci. USA* **74**, 3642.

Appendix 1

Important *Escherichia coli* strains

T. A. Brown

Department of Biomolecular Sciences, University of Manchester Institute of Science and Technology, Manchester M60 IQD, UK

1 Genotypes of important strains

A comprehensive listing of the genotypes of many *E. coli* strains used in molecular biology research is given in ref. 1. This source also provides a detailed explanation of the nomenclature used when writing out a genotype as well information on the restriction and modification properties of important strains.

1.1 Strains for general purpose cloning

The following four strains are popular hosts for cloning vectors that do not employ Lac selection.

C600   *supE44 thi-1 thr-1 leuB6 lacY1 tonA21 mcrA*

DH1   *supE44 hsdR17 recA1 endA1 gyrA96 thi-1 relA1*

HB101 Δ*(gpt-proA)62 leuB6 supE44 ara-14 galK2 lacY1 rpsL20 xyl-5* Δ*(mcrCB-hsdSMR-mrr) mtl-1 recA13*

RR1   Δ*(gpt-proA)62 leuB6 supE44 ara-14 galK2 lacY1 rpsL20 xyl-5* Δ*(mcrCB-hsdSMR-mrr) mtl-1*

These four strains have been used in research for many years and versions with slightly different genotypes might be in circulation (1). For example, some strains of C600 are *mcrA*$^+$ and DH1 may also be *spoT1 rfbD1*. HB101 and RR1 might be *leu*$^+$ and/or *thi-1*.

1.2 Strains for Lac selection

The following strains can be used for general purpose cloning and also with plasmid and M13 vectors that utilize Lac selection. MV1184 and XL1-Blue are popular hosts for the generation of single-stranded DNA from phagemid vectors, using M13KO7 as the helper phage. The presence of the *recB recJ sbcC umuC uvrC* mutations in SURE helps to prevent rearrangement of cloned DNA sequences. SURE, XL1-Blue and XL1-Blue MRF′ are marketed by Stratagene; all three strains carry undescribed mutations that give colonies a more intensely blue coloration on agar plates containing X-gal and IPTG.

JM83   *ara Δ(lac-proAB) rpsL (φ80 lacZΔM15)*

JM109   *recA1 supE44 endA1 hsdR17 gyrA96 relA1 thi-1 mcrA Δ(lac-proAB)*
        F′[traD36 proAB⁺ lacIq lacZΔM15]

JM110   *dam dcm supE44 thi-1 leu rpsL lacY galK galT ara tonA thr tsx Δ(lac-proAB)*
        F′[traD36 proAB⁺ lacIq lacZΔM15]

KK2186 *supE thi-1 endA1 hsdR4 sbcBC rpsL Δ(lac-proAB) F′[traD36 proAB⁺ lacIq*
        lacZΔM15]

MV1184 *Δ(lac-proAB) rpsL thi (φ80 lacZΔM15) Δ(srl-recA)306::Tn10(tetr)*
        F′[traD36 proAB⁺ lacIq lacZΔM15]

SURE   *mcrA Δ(mcrCB-hsdSMR-mrr)171 endA1 supE44 thi-1 gyrA96 relA1 lac*
        recB recJ sbcC umuC::Tn5(kanr) uvrC F′[proAB⁺ lacIq lacZΔM15 Tn10(tetr)]

XL1-Blue   *recA1 endA1 gyrA96 thi-1 hsdR17 supE44 relA1 lac F′[proAB⁺ lacIq*
            lacZΔM15 Tn10(tetr)]

XL1-Blue MRF′   *Δ(mcrA)183 Δ(mcrCB-hsdSMR-mrr)173 recA1 endA1 gyrA96 thi-1 supE44*
                relA1 lac F′[proAB⁺ lacIq lacZΔM15 Tn5(kanr)]

The genotype of JM110 is described by some sources as also being *hsdR17* but this is probably incorrect. KK2186 is identical to the popular strain JM103, but KK2186 should be used instead of JM103 because some examples of the latter have accumulated additional mutations.

1.3 Strains for protein expression with pET vectors

The expression of protein from genes cloned in pET vectors requires that the host cells possess the gene for the T7 RNA polymerase (Chapter 9, Section 3.5). This gene is carried by an engineered version of bacteriophage λ called λDE3 and the relevant strains contain this λ genome as a lysogen:

BL21(DE3)   *hsdS gal (λcIts857 ind1 Sam7 nin5 lacUV5-T7 gene 1)*

JM109(DE3) *recA1 supE44 endA1 hsdR17 gyrA96 relA1 thi-1 mcrA Δ(lac-proAB)*
            F′[traD36 proAB⁺ lacIq lacZΔM15] (λcIts857 ind1 Sam7 nin5 lacUV5-T7
            gene 1)

TKB1   *dcm ompT hsdS gal (λcIts857 ind1 Sam7 nin5 lacUV5-T7 gene 1) (pTK*
        tetr)

1.4 Strains for λ vectors

The following strains are frequently used with bacteriophage λ vectors.

K802   *supE44 hsdR galK2 galT22 metB1 mcrA mcrB1*

NM646   *supE44 hsdR galK2 galT22 metB1 mcrA mcrB1 (P2cox3)*

Y1088   *Δ(lac)U169 supE supF hsdR metB trpR tonA21 mcrA proC::Tn5(kanr) (pMC9)*

Y1090   *mcrA araD139 Δ(lac)U169 lon-100 rpsL supF trpC22::Tn10(tetr) (pMC9)*

In the Y1088 and Y1090 genotypes '(pMC9)' denotes possession of pBR322 carrying *lacI^q^*.

1.5 Strains for λ packaging extracts

The strains used as sources of the λ packaging extracts (Chapter 8, *Protocol 6*) have the following genotypes.

BHB2688 N205 *recA* (λ *imm*434 *c*Its *b2 red3* Eam4 Sam7)/λ

BHB2690 N205 *recA* (λ *imm*434 *c*Its *b2 red3* Dam15 Sam7)/λ

2 Key features of *E. coli* genotypes

2.1 Recombination deficiency

The major recombination pathways in *E. coli* are coded by the *rec* genes. Strains that are Rec$^+$ are less suitable for cloning because they can multimerize plasmids and rearrange inserted DNA (2, 3). There are at least nine different *rec* genes, with host strains for molecular biology experiments often carrying one or more of *recA*, *recBC*, *recD*, *recF* and *recJ*. The mutation *sbcBC* may also be present to further reduce rearrangement of cloned DNA (4, 5). One problem with *rec* strains is that they are relatively slow growing and may be difficult to transform. Strains used as hosts for bacteriophage λ vectors that are *red gam* (Chapter 9, Section 4.1) must be RecA$^+$.

2.2 Host-controlled restriction-modification

There are four different *E. coli* restriction and/or modification systems that are relevant to the use of this bacterium in gene cloning (1):

- Most strains used for cloning are derived from *E. coli* K, which possesses the *EcoK* restriction-modification system (6). This is coded by the *hsd* genes, with an *hsdM* strain being methylation-deficient, *hsdR* being restriction-deficient and *hsdS* lacking both functions. Strains used in cloning are usually *hsdR* or *hsdS* so that there is no possibility of *EcoK* restriction sites in the cloned DNA being cleaved.

- Many host strains for cloning are also *mcrA* and *mcrBC* and so lack functional McrA and McrB restriction systems, which cleave DNA at short target sequences and which, if not mutated, could cleaved cloned DNA (6, 7–9).

- Similarly, many strains are *mrr* and so lack the Mrr restriction system (6, 8, 10). Mrr is similar to McrA and McrBC, although the identity of the restriction sequence is not known.

- The DNA adenine methylase and DNA cytosine methylase add methyl groups to target sequences (6, 11–13). Strains used for cloning are sometimes *dam* and/or *dcm* so that the cloned DNA does not become modified. This avoids possible problems that might arise after the cloned DNA is purified, because the methylated DNA may not be cut by a restriction enzyme if the recognition sequence for the enzyme overlaps a methylation site and the enzyme action is inhibited by the presence of the methyl group.

Comprehensive details of the restriction-modification properties of *E. coli* strains are given in ref. 1.

2.3 Suppressor mutations

The amber suppressor mutations *supE* and *supF* result in insertion of glutamine and tyrosine, respectively, at UAG codons. These mutations provide a form of biological containment in that phages carrying amber mutations can survive only in a suppressing strain and so cannot escape to natural populations of bacteria.

2.4 Genotypes relevant to Lac selection

Plasmids carrying the *lacZ'* gene require special hosts. These usually carry a chromosomal deletion called Δ(*lac-proAB*), which spans the lactose operon and surrounding region. The deletion is partially complemented by an engineered F' plasmid, which carries *proAB*⁺ (rescuing proline auxotrophy), and *lacZΔM15*, which is the *lac* operon minus the N-terminal α-peptide of *lacZ* (amino acids 11–41) which is the part present in the *lacZ'* segment carried by the vector. The F' plasmid also carries *lacI*^q, a mutation that results in over-expression of the *lac* repressor (14) and reduces the background expression of the *lac* operon that occurs in uninduced cells. The presence of this mutation therefore makes the X-gal identification system more discriminatory by ensuring that non-recombinant cells are cream and not pale blue.

The genes described above provide all of the sequence information needed for utilization of lactose except for the α-peptide coded by *lacZ'*. The presence of the *proAB*⁺ genes on the F' plasmid means that retention of the F' can be selected for by maintaining the bacteria on a proline-deficient minimal medium. This is important because F' bacteria are unstable and can revert to F⁻ if stored for long periods.

References

1. Brown, T. A. (ed.) (1998). *Molecular biology labfax*, 2nd edn, Vol. 1. Academic Press, London.
2. Yanisch-Perron, C., Vieira, J., and Messing, J. (1985). *Gene* **33**, 103.
3. Bedbrook, J. R. and Ausubel, F. M. (1976). *Cell* **9**, 707.
4. Kushner, S. R., Nagaishi, H., Templin, A., and Clark, A. J. (1971). *Proc. Natl. Acad. Sci. USA* **75**, 2276.
5. Leach, D. R. F. and Stahl, F. W. (1983). *Nature* **305**, 448.
6. Raleigh, E. A., Lech, K. and Brent, R. (1989). In *Current protocols in molecular biology* (ed. F. M. Ausubel, R. Brent, R. E. Kingston, D. D. Moore, J. G. Seidman, J. A. Smith, and K. Struhl), p. 1.4.6. John Wiley, New York.
7. Raleigh, E. A. (1992). *Mol. Microbiol.* **6**, 1079.
8. Kelleher, J. and Raleigh, E. A. (1991). *J. Bacteriol.* **173**, 5220.

9. Sutherland, E., Coe, L., and Raleigh, E. A. (1992). *J. Mol. Biol.* **225**, 327.

10. Waite-Rees, P. A., Keating C. J., Moran, L. S., Stalko, B. E., Hornstra, L. J., and Benner J. S. (1991). *J. Bacteriol.* **173**, 5207.

11. Hattman, S., Brooks, J. E., and Masurekar, M. (1978). *J. Mol. Biol.*, **126**, 367.

12. Marinus, M. G. and Morris, N. R. (1973). *J. Bacteriol.* **114**, 1143.

13. May, M. S. and Hattman, S. (1975). *J. Bacteriol.* **123**, 768.

14. Muller-Hill, B., Crapo, L., and Gilbert, W. (1968). *Proc. Natl. Acad. Sci. USA* **59**, 1259.

Appendix 2
Recipes and general procedures

T. A. Brown

Department of Biomolecular Sciences, University of Manchester Institute of
Science and Technology, Manchester M60 IQD, UK

1 Recipes for *Escherichia coli* culture media

1.1 Liquid media

Recipes for 1 litre of media

BBL broth	18 g BBL trypticase peptone, 5 g NaCl
DYT	16 g bacto-tryptone, 10 g bacto-yeast extract, 5 g NaCl
LB	10 g bacto-tryptone, 5 g bacto-yeast extract, 10 g NaCl
Nutrient broth	25 g bacto-nutrient broth
NZY	10 g NZ amine, 5 g bacto-yeast extract, 2 g $MgSO_4$
SOB	20 g bacto-tryptone, 5 g bacto-yeast extract, 0.5 g NaCl
Superbroth (SB)	32 g bacto-tryptone, 20 g bacto-yeast extract, 5 g NaCl, 5 ml 1 M NaOH
Terrific broth (TB)	12 g bacto-tryptone, 24 g bacto-yeast extract, 4 ml glycerol. Prepare in 850 ml water, autoclave, cool, add 100 ml of a sterile solution of 0.17 M KH_2PO_4 + 0.72 M K_2HPO_4 and make the final volume up to 1 l with sterile water
Tryptone broth	10 g bacto-tryptone, 5 g NaCl
YT	8 g bacto-tryptone, 5 g bacto-yeast extract, 5 g NaCl

With all media the pH should be checked and if necessary adjusted to 7.0–7.2
with NaOH. Media should be sterilized by autoclaving at 121 °C, 103.5 kPa (15
lb/in²) for 20 min.

1.2 Solid media

To prepare solid media add the required amount of bacto-agar before autoclaving. For agar plates add 15 g/l.

1.2.1 Soft agar overlays

To prepare soft or top agar overlays use the recipes given above but add a reduced
amount of agar or agarose:

- LTS is DYT plus 6 g agar/l

- NZY top agar is NZY plus 7 g agarose/l
- YTS is YT plus 6 g agar/l

1.3 Supplements

1.3.1 Supplements for λ phage

Maltose Prepare a 20% (w/v) stock solution and filter-sterilize. Store at room temperature and add to autoclaved media (final concentration 0.2%) immediately before use.

MgSO$_4$ Prepare a 1 M stock solution, autoclave and add to autoclaved media (final concentration usually 10 mM) immediately before use.

1.3.2 Antibiotics

All antibiotics must be prepared as filter-sterilized stock solutions and added to autoclaved media immediately before use. Agar media must be cooled to 50°C before antibiotic supplementation as many are heat-sensitive.

Ampicillin
- stock: 50 mg/ml in water
- working concentration: 40–60 μg/ml (plates); 25 μg/ml (broth)

Kanamycin
- stock: 35 mg/ml in water
- working concentration: 25–50 μg/ml (plates); 25–70 μg/ml (broth)

Streptomycin
- stock: 10 mg/ml in water
- working concentration: 40 μg/ml (plates); 15 μg/ml (broth)

Tetracycline
- stock: 5 mg/ml in ethanol
- working concentration: 15 μg/ml (plates); 7.5 μg/ml (broth)

Note that tetracycline is light-sensitive and all solutions and plates should be wrapped in foil or stored in the dark. Magnesium ions antagonize the activity of tetracycline.

2 Buffers

PBS: 8.0 g NaCl, 0.34 g KH$_2$PO$_4$, 1.21 g K$_2$HPO$_4$ per 1 l; pH should be 7.3. Sterilize by autoclaving.

SM (λ storage and dilution): 5.8 g NaCl, 2 g MgSO$_4$.7H$_2$O, 50 ml 1 M Tris-HCl, pH 7.5, 5 ml 2% (w/v) gelatin solution, water to make 1 l. Sterilize by autoclaving.

TE buffers: 10 mM Tris-HCl, 1 mM EDTA, pH 8.0. The pH of the Tris-HCl determines the pH of the TE buffer.

3 General procedures

Protocols 1–5 describe five general procedures required as support for many different molecular biology procedures.

Protocol 1

Preparation of DNase-free RNase A[a]

Reagents

- Ribonuclease A (RNase A)

Method

1 Dissolve solid RNase A to 10 mg/ml in 10 mM Tris-HCl, pH 7.5, 15 mM NaCl.

2 Heat to 100°C for 15 min.

3 Cool slowly to room temperature. Store in aliquots at −20°C.

[a] Commercial supplies of RNase A usually contain DNase activity.

Protocol 2

Preparation of siliconized glassware

Reagents

- Siliconizing solution (2% dimethyldichlorosilane in 1,1,1-trichloroethane) (Merck)

Method

1 Rinse the glassware in siliconizing solution. The solution can be reused several times. Avoid contact with the skin. Use a fume hood.

2 Allow the glassware to dry thoroughly (up to 2 h).

3 Rinse each item twice in double-distilled water.

Notes

(a) DNA will adsorb to non-siliconized glassware, resulting in substantial losses. Glass wool can be treated in exactly the same way. Plasticware does not usually need to be siliconized, although some applications require it.

(b) If a desiccator can be dedicated to siliconization then instead of step 1 evaporate 1 ml of siliconizing solution under vacuum inside the desiccator with the glassware to be treated.

(c) **Caution: siliconizing solutions are toxic and flammable.**

Protocol 3

Preparation of non-homologous DNA for hybridization analysis

Equipment and Reagents

- Salmon sperm DNA (Sigma Type III, sodium salt)
- 17-G syringe needle

Method

1 Prepare a 10 mg/ml solution of salmon sperm DNA in water.

2 Adjust the sodium concentration to 0.1 M and extract once with phenol and once with 25:24:1 (v/v/v) phenol–chloroform–isoamyl alcohol. This step can be omitted with most batches of salmon sperm DNA but should be included if poor results (i.e. high background hybridization) are obtained.

3 Pass through a 17-G syringe needle approximately 20 times. This should be done in a vigorous manner as the aim is to shear the DNA.

4 Add 2 volumes of cold absolute ethanol, leave on ice for 10 min, then spin in the microfuge for 30 min. Remove the ethanol, dry the pellet, and resuspend in the original volume of water. Again, this step may be unnecessary.

5 Place in a boiling water bath for 10 min, chill and then either:

   (a) Use immediately at a final concentration of 100 µg/ml.

   (b) Store at −20°C in small aliquots. Immediately before use place in a boiling water bath for 5 min and chill on ice.

Note

Non-homologous DNA is often added to pre-hybridization and hybridization solutions to block non-specific binding sites on the filter surface (see Volume II, Chapter 5). The non-homologous DNA is usually obtained from salmon sperm, herring sperm or calf thymus.

Protocol 4

Preparation of deionized formamide

Reagents

- Formamide
- Whatman No. 1 paper
- Mixed bed ion-exchange resin (e.g. Bio-Rad AG 501-X8 or X8(D) resins)

Method

1 Add 5 g of mixed bed ion-exchange resin per 100 ml formamide.

2 Stir at room temperature for 1 h.

Protocol 4 continued

3 Filter twice through Whatman No. 1 paper.

Notes

(a) Commercial formamide is usually satisfactory for use in molecular biology experiments (e.g. in hybridization solutions and loading buffers for gel electrophoresis) but if the liquid has a yellow colour it should be deionized until it becomes colourless, as described in this Protocol.

(b) The resin can be reused a number of times. The X8(D) resin contains a dye that indicates when it is exhausted.

Protocol 5

Decontamination of ethidium bromide solutions

Reagents

• Amberlite XAD-16 resin (Bio-Rad)

Method

1 Dilute the ethidium bromide solution to a concentration no greater than 100 µg/ml.

2 Add 0.5 g of Amberlite XAD-16 per 10 ml ethidium bromide solution.

3 Stir for 2 h for every 10 ml of solution (e.g. stir a 20 ml solution for 4 h and a 100 ml solution for 20 h).

4 Filter the mixture to collect the resin. The filtrate can be treated as non-hazardous waste and disposed of in the normal way. The resin must be treated as hazardous waste.

Appendix 3
Restriction endonucleases

T. A. Brown

Department of Biomolecular Sciences, University of Manchester Institute of Science and Technology, Manchester M60 1QD, UK

1 Details of restriction endonucleases

Numerous restriction endonucleases recognizing a variety of target sequences have been identified. A complete and up-to-date listing is provided in ref. 1. Extensive details of reactions conditions for different enzymes are given in ref. 2.

1.1 Recognition sequences of commonly–used enzymes

Table 1 gives the recognition sequences for many of the restriction endonucleases commonly used in molecular biology research. It is based on ref. 2, which contains a more comprehensive listing. Consult ref. 2 for the recognition sequences of enzymes not listed here or in ref. 1.

Table 1 Recognition sequences for commonly-used restriction endonucleases[a]

Enzyme	Recognition sequence	Type of end	Sequence of overhang
*Aat*I	AGG ↑ CCT	Blunt	
*Aat*II	GACGT ↑ C	3′-overhang	–ACGT
*Acc*I	GT ↑ MKAC	5′-overhang	MK–
*Acc*II	CG ↑ CG	Blunt	
*Acc*III	T ↑ CCGGA	5′-overhang	CCGG–
*Aci*I	CCGC(–3/–1)	5′-overhang	NN–
*Acs*I	R ↑ AATTY	5′-overhang	AATT–
*Acy*I	GR ↑ CGYC	5′-overhang	CG–
*Afa*I	GT ↑ AC	Blunt	
*Afl*I	G ↑ GWCC	5′-overhang	GWC–
*Afl*II	C ↑ TTAAG	5′-overhang	TTAA–
*Afl*III	A ↑ CRYGT	5′-overhang	CRYG–
*Age*I	A ↑ CCGGT	5′-overhang	CCGG–
*Aha*I	CC ↑ SGG	5′-overhang	S–
*Aha*II	GR ↑ CGYC	5′-overhang	CG–
*Aha*III	TTT ↑ AAA	Blunt	

221

Table 1 Continued

Enzyme	Recognition sequence	Type of end	Sequence of overhang
*Ahd*I	GACNNN ↑ NNGTC	3'-overhang	–N
*Alu*I	AG ↑ CT	Blunt	
*Alw*I	GGATC(4/5)	5'-overhang	N–
*Alw*NI	CAGNNN ↑ CTG	3'-overhang	–NNN
*Aoc*I	CC ↑ TNAGG	5'-overhang	TNA–
*Aoc*II	GDGCH ↑ C	3'-overhang	–DGCH
*Aos*I	TGC ↑ GCA	Blunt	
*Apa*I	GGGCC ↑ C	3'-overhang	–GGCC
*Apa*LI	G ↑ TGCAC	5'-overhang	TGCA–
*Apo*I	R ↑ AATTY	5'-overhang	AATT–
*Apy*I	CC ↑ WGG	5'-overhang	W–
*Aqu*I	C ↑ YCGRG	5'-overhang	YCGR–
*Asc*I	GG ↑ CGCGCC	5'-overhang	CGCG–
*Ase*I	AT ↑ TAAT	5'-overhang	TA–
*Asn*I	AT ↑ TAAT	5'-overhang	TA–
*Asp*I	GACN ↑ NNGTC	5'-overhang	N–
*Asp*700I	GAANN ↑ NNTTC	Blunt	
*Asp*EI	GACNNN ↑ NNGTC	3'-overhang	–N
*Asp*HI	GWGCW ↑ C	3'-overhang	–WGCW
*Asu*I	G ↑ GNCC	5'-overhang	GNC–
*Asu*II	TT ↑ CGAA	5'-overhang	CG–
*Ava*I	C ↑ YCGRG	5'-overhang	YCGR–
*Ava*II	G ↑ GWCC	5'-overhang	GWC–
*Avi*II	TGC ↑ GCA	Blunt	
*Avr*II	C ↑ CTAGG	5'-overhang	CTAG–
*Axy*I	CC ↑ TNAGG	5'-overhang	TNA–
*Bal*I	TGG ↑ CCA	Blunt	
*Bam*HI	G ↑ GATCC	5'-overhang	GATC–
*Ban*I	G ↑ GYRCC	5'-overhang	GYRC–
*Ban*II	GRGCY ↑ C	3'-overhang	–RGCY
*Ban*III	AT ↑ CGAT	5'-overhang	CG–
*Bbe*I	GGCGC ↑ C	3'-overhang	–GCGC
*Bbi*II	GR ↑ CGYC	5'-overhang	CG–
*Bbr*PI	CAC ↑ GTG	Blunt	
*Bbs*I	GAAGAC(2/6)	5'-overhang	NNNN–
*Bbu*I	GCATG ↑ C	3'-overhang	–CATG
*Bbv*I	GCAGC(8/12)	5'-overhang	NNNN–
*Bcg*I	(10/12)GCANNNNNNTCG(12/10)[b]		
*Bcl*I	T ↑ GATCA	5'-overhang	GATC–
*Bcn*I	CC ↑ SGG	5'-overhang	S–
*Bfa*I	C ↑ TAG	5'-overhang	TA–

Table 1 Continued

Enzyme	Recognition sequence	Type of end	Sequence of overhang
*Bfr*I	C ↑ TTAAG	5′-overhang	TTAA–
*Bgl*I	GCCNNNN ↑ NGGC	3′-overhang	–NNN
*Bgl*II	A ↑ GATCT	5′-overhang	GATC–
*Bln*I	C ↑ CTAGG	5′-overhang	CTAG–
*Blp*I	GC ↑ TNAGC	5′-overhang	TNA–
*Bmy*I	GDGCH ↑ C	3′-overhang	–DGCH
*Bpm*I	CTGGAG(16/14)	3′-overhang	–NN
*Bpu*AI	GAAGAC(2/6)	5′-overhang	NNNN–
*Bsa*I	GGTCTC(1/5)	5′-overhang	NNNN–
*Bsa*AI	YAC ↑ GTR	Blunt	
*Bsa*BI	GATNN ↑ NNATC	Blunt	
*Bsa*HI	GR ↑ CGYC	5′-overhang	CG–
*Bsa*JI	C ↑ CNNGG	5′-overhang	CNNG–
*Bsa*MI	GAATGC(1/−1)	3′-overhang	–NGAATGCN
*Bsa*OI	CGRY ↑ CG	3′-overhang	–RY
*Bsa*WI	W ↑ CCGGW	5′-overhang	CCGG–
*Bse*AI	T ↑ CCGGA	5′-overhang	CCGG–
*Bse*RI	GAGGAG(10/8)	3′-overhang	–NN
*Bsg*I	GTGCAG(16/14)	3′-overhang	–NNNN
*Bsi*CI	TT ↑ CGAA	5′-overhang	CG–
*Bsi*EI	CGRY ↑ CG	3′-overhang	–RY
*Bsi*WI	C ↑ GTACG	5′-overhang	GTAC–
*Bsi*YI	CCNNNNN ↑ NNGG	3′-overhang	–NNN
*Bsl*I	CCNNNNN ↑ NNGG	3′-overhang	–NNN
*Bsm*I	GAATGC(1/−1)	3′-overhang	–NGAATGCN
*Bsm*AI	GTCTC(1/5)	5′-overhang	NNNN–
*Bsm*BI	CGTCTC(1/5)	5′-overhang	NNNN–
*Bsm*FI	GGGAC(10/14)	5′-overhang	NNNN–
*Bso*BI	C ↑ YCGRG	5′-overhang	YCGR–
*Bsp*1286I	GDGCH ↑ C	3′-overhang	–DGCH
*Bsp*CI	CGAT ↑ CG	3′-overhang	–T
*Bsp*DI	AT ↑ CGAT	5′-overhang	CG–
*Bsp*EI	T ↑ CCGGA	5′-overhang	CCGG–
*Bsp*HI	T ↑ CATGA	5′-overhang	CATG–
*Bsp*MI	ACCTGC(4/8)	5′-overhang	NNNN–
*Bsp*MII	T ↑ CCGGA	5′-overhang	CCGG–
*Bsr*I	ACTGG(1/−1)	3′-overhang	–NACTGGN
*Bsr*BI	CCGCTC(−3/−3)	Blunt	
*Bsr*BRI	GATNN ↑ NNATC	Blunt	
*Bsr*DI	GCAATG(2/0)	3′-overhang	–NN
*Bsr*FI	R ↑ CCGGY	5′-overhang	CCGG–

Table 1 Continued

Enzyme	Recognition sequence	Type of end	Sequence of overhang
*Bsr*GI	T ↑ GTACA	5'-overhang	GTAC–
*Bsr*SI	ACTGG(1/–1)	3'-overhang	–NACTGGN
*Bss*HII	G ↑ CGCGC	5'-overhang	CGCG–
*Bss*KI	↑ CCNGG	5'-overhang	CCNGG–
*Bss*SI	CACGAG(–5/–1)	5'-overhang	NNNN–
*Bst*I	G ↑ GATCC	5'-overhang	GATC–
*Bst*BI	TT ↑ CGAA	5'-overhang	CG–
*Bst*EII	G ↑ GTNACC	5'-overhang	GTNAC–
*Bst*NI	CC ↑ WGG	5'-overhang	W–
*Bst*OI	CC ↑ WGG	5'-overhang	W–
*Bst*PI	G ↑ GTNACC	5'-overhang	GTNAC–
*Bst*UI	CG ↑ CG	Blunt	
*Bst*XI	CCANNNNN ↑ NTGG	3'-overhang	–NNNN
*Bst*YI	R ↑ GATCY	5'-overhang	GATC–
*Bst*ZI	C ↑ GGCCG	5'-overhang	GGCC–
*Cac*8I	GCN ↑ NGC	Blunt	
*Cfo*I	GCG ↑ C	3'-overhang	–CG
*Cfr*9I	C ↑ CCGGG	5'-overhang	CCGG–
*Cfr*10I	R ↑ CCGGY	5'-overhang	CCGG–
*Cfr*13I	G ↑ GNCC	5'-overhang	GNC–
*Cla*I	AT ↑ CGAT	5'-overhang	CG–
*Cpo*I	CG ↑ GWCCG	5'-overhang	GWC–
*Csp*I	CG ↑ GWCCG	5'-overhang	GWC–
*Csp*45I	TT ↑ CGAA	5'-overhang	CG–
*Cvn*I	CC ↑ TNAGG	5'-overhang	TNA–
*Dde*I	C ↑ TNAG	5'-overhang	TNA–
*Dpn*I	GA ↑ TC	Blunt	
*Dpn*II	↑ GATC	5'-overhang	GATC–
*Dra*I	TTT ↑ AAA	Blunt	
*Dra*II	RG ↑ GNCCY	5'-overhang	GNC–
*Dra*III	CACNNN ↑ GTG	3'-overhang	–NNN
*Drd*I	GACNNNN ↑ NNGTC	3'-overhang	–NN
*Eae*I	Y ↑ GGCCR	5'-overhang	GGCC–
*Eag*I	C ↑ GGCCG	5'-overhang	GGCC–
*Eam*1105I	GACNNN ↑ NNGTC	3'-overhang	–N
*Ear*I	CTCTTC(1/4)	5'-overhang	NNN–
*Eco*NI	CCTNN ↑ NNNAGG	5'-overhang	N–
*Eco*RI	G ↑ AATTC	5'-overhang	AATT–
*Eco*RII	↑ CCWGG	5'-overhang	CCWGG–
*Eco*RV	GAT ↑ ATC	Blunt	
*Ehe*I	GGC ↑ GCC	Blunt	

Table 1 Continued

Enzyme	Recognition sequence	Type of end	Sequence of overhang
*Esp*I	GC ↑ TNAGC	5′-overhang	TNA–
*Fba*I	T ↑ GATCA	5′-overhang	GATC–
*Fdi*II	TGC ↑ GCA	Blunt	
*Fnu*DII	CG ↑ CG	Blunt	
*Fok*I	GGATG(9/13)	5′-overhang	NNNN–
*Fse*I	GGCCGG ↑ CC	3′-overhang	–CCGG
*Fsp*I	TGC ↑ GCA	Blunt	
*Hae*II	RGCGC ↑ Y	3′-overhang	–GCGC
*Hae*III	GG ↑ CC	Blunt	
*Hap*II	C ↑ CGG	5′-overhang	CG–
*Hga*I	GACGC(5/10)	5′-overhang	NNNNN–
*Hgi*AI	GWGCW ↑ C	3′-overhang	–WGCW
*Hha*I	GCG ↑ C	3′-overhang	–CG
*Hin*1I	GR ↑ CGYC	5′-overhang	CG–
*Hin*cII	GTY ↑ RAC	Blunt	
*Hin*dII	GTY ↑ RAC	Blunt	
*Hin*dIII	A ↑ AGCTT	5′-overhang	AGCT–
*Hin*fI	G ↑ ANTC	5′-overhang	ANT–
*Hin*P1I	G ↑ CGC	5′-overhang	CG–
*Hpa*I	GTT ↑ AAC	Blunt	
*Hpa*II	C ↑ CGG	5′-overhang	CG–
*Hph*I	GGTGA(8/7)	3′-overhang	–N
*Ita*I	GC ↑ NGC	5′-overhang	N–
*Kas*I	G ↑ GCGCC	5′-overhang	GCGC–
*Kpn*I	GGTAC ↑ C	3′-overhang	–GTAC
*Ksp*I	CCGC ↑ GG	3′-overhang	–GC
*Mae*I	C ↑ TAG	5′-overhang	TA–
*Mae*II	A ↑ CGT	5′-overhang	CG–
*Mae*III	↑ GTNAC	5′-overhang	GTNAC–
*Mam*I	GATNN ↑ NNATC	Blunt	
*Mbo*I	↑ GATC	5′-overhang	GATC–
*Mbo*II	GAAGA(8/7)	3′-overhang	–N
*Mfe*I	C ↑ AATTG	5′-overhang	AATT–
*Mfl*I	R ↑ GATCY	5′-overhang	GATC–
*Mlu*I	A ↑ CGCGT	5′-overhang	CGCG–
*Mlu*NI	TGG ↑ CCA	Blunt	
*Mnl*I	CCTC(7/6)	3′-overhang	–N
*Mro*I	T ↑ CCGGA	5′-overhang	CCGG–
*Msc*I	TGG ↑ CCA	Blunt	
*Mse*I	T ↑ TAA	5′-overhang	TA–
*Msl*I	CAYNN ↑ NNRTG	Blunt	

Table 1 Continued

Enzyme	Recognition sequence	Type of end	Sequence of overhang
*Msp*I	C ↑ CGG	5'-overhang	CG–
*Msp*A1I	CMG ↑ CKG	Blunt	
*Mst*I	TGC ↑ GCA	Blunt	
*Mst*II	CC ↑ TNAGG	5'-overhang	TNA–
*Mun*I	C ↑ AATTG	5'-overhang	AATT–
*Mva*I	CC ↑ WGG	5'-overhang	W–
*Mvn*I	CG ↑ CG	Blunt	
*Mwo*I	GCNNNNN ↑ NNGC	3'-overhang	–NNN
*Nae*I	GCC ↑ GGC	Blunt	
*Nar*I	GG ↑ CGCC	5'-overhang	CG–
*Nci*I	CC ↑ SGG	5'-overhang	S–
*Nco*I	C ↑ CATGG	5'-overhang	CATG–
*Nde*I	CA ↑ TATG	5'-overhang	TA–
*Nde*II	↑ GATC	5'-overhang	GATC–
*Ngo*MI	G ↑ CCGGC	5'-overhang	CCGG–
*Nhe*I	G ↑ CTAGC	5'-overhang	CTAG–
*Nla*III	CATG ↑	3'-overhang	–CATG
*Nla*IV	GGN ↑ NCC	Blunt	
*Not*I	GC ↑ GGCCGC	5'-overhang	GGCC–
*Nru*I	TCG ↑ CGA	Blunt	
*Nsi*I	ATGCA ↑ T	3'-overhang	–TGCA
*Nsp*I	RCATG ↑ Y	3'-overhang	–CATG
*Nsp*II	GDGCH ↑ C	3'-overhang	–DGCH
*Nsp*III	C ↑ YCGRG	5'-overhang	YCGR–
*Nsp*IV	G ↑ GNCC	5'-overhang	GNC–
*Nsp*V	TT ↑ CGAA	5'-overhang	CG–
*Nsp*BII	CMG ↑ CKG	Blunt	
*Nsp*HI	RCATG ↑ Y	3'-overhang	–CATG
*Nun*II	GG ↑ CGCC	5'-overhang	CG–
*Pac*I	TTAAT ↑ TAA	3'-overhang	–AT
*Pae*R7I	C ↑ TCGAG	5'-overhang	TCGA–
*Pal*I	GG ↑ CC	Blunt	
*Pfl*FI	GACN ↑ NNGTC	5'-overhang	N–
*Pfl*MI	CCANNNN ↑ NTGG	3'-overhang	–NNN
*Ple*I	GAGTC(4/5)	5'-overhang	N–
*Pma*CI	CAC ↑ GTG	Blunt	
*Pme*I	GTTT ↑ AAAC	Blunt	
*Pml*I	CAC ↑ GTG	Blunt	
*Ppu*MI	RG ↑ GWCCY	5'-overhang	GWC–
*Psh*AI	GACNN ↑ NNGTC	Blunt	
*Psp*AI	C ↑ CCGGG	5'-overhang	CCGG–

Table 1 Continued

Enzyme	Recognition sequence	Type of end	Sequence of overhang
*Pss*I	RGGNC ↑ CY	3′-overhang	–GNC
*Pst*I	CTGCA ↑ G	3′-overhang	–TGCA
*Pvu*I	CGAT ↑ CG	3′-overhang	–AT
*Pvu*II	CAG ↑ CTG	Blunt	
*Rsa*I	GT ↑ AC	Blunt	
*Rsp*XI	T ↑ CATGA	5′-overhang	CATG–
*Rsr*I	G ↑ AATTC	5′-overhang	AATT–
*Rsr*II	CG ↑ GWCCG	5′-overhang	GWC–
*Sac*I	GAGCT ↑ C	3′-overhang	–AGCT
*Sac*II	CCGC ↑ GG	3′-overhang	–GC
*Sal*I	G ↑ TCGAC	5′-overhang	TCGA–
*San*DI	GG ↑ GWCCC	5′-overhang	GWC–
*Sap*I	GCTCTTC(1/4)	5′-overhang	NNN–
*Sau*I	CC ↑ TNAGG	5′-overhang	TNA–
*Sau*3AI	↑ GATC	5′-overhang	GATC–
*Sau*96I	G ↑ GNCC	5′-overhang	GNC–
*Sca*I	AGT ↑ ACT	Blunt	
*Scr*FI	CC ↑ NGG	5′-overhang	N–
*Sdu*I	GDGCH ↑ C	3′-overhang	–DGCH
*Sex*AI	A ↑ CCWGGT	5′-overhang	CCWGG–
*Sfa*NI	GCATC(5/9)	5′-overhang	NNNN–
*Sfc*I	C ↑ TRYAG	5′-overhang	TRYA–
*Sfi*I	GGCCNNNN ↑ NGGCC	3′-overhang	–NNN
*Sfu*I	TT ↑ CGAA	5′-overhang	CG–
*Sgf*I	GCGAT ↑ CGC	3′-overhang	–AT
*Sgr*AI	CR ↑ CCGGYG	5′-overhang	CCGG–
*Sin*I	G ↑ GWCC	5′-overhang	GWC–
*Sma*I	CCC ↑ GGG	Blunt	
*Sml*I	C ↑ TYRAG	5′-overhang	TY–
*Sna*BI	TAC ↑ GTA	Blunt	
*Spe*I	A ↑ CTAGT	5′-overhang	CTAG–
*Sph*I	GCATG ↑ C	3′-overhang	–CATG
*Spl*I	C ↑ GTACG	5′-overhang	GTAC–
*Spo*I	TCG ↑ CGA	Blunt	
*Srf*I	GCCC ↑ GGGC	Blunt	
*Ssp*I	AAT ↑ ATT	Blunt	
*Ssp*BI	T ↑ GTACA	5′-overhang	GTAC–
*Sst*I	GAGCT ↑ C	3′-overhang	–AGCT
*Sst*II	CCGC ↑ GG	3′-overhang	–GC
*Stu*I	AGG ↑ CCT	Blunt	
*Sty*I	C ↑ CWWGG	5′-overhang	CWWG–

227

Table 1 Continued

Enzyme	Recognition sequence	Type of end	Sequence of overhang
SwaI	ATTT ↑ AAAT	Blunt	
TaiI	ACGT ↑	3′-overhang	–ACGT
TaqI	T ↑ CGA	5′-overhang	CG–
TfiI	G ↑ AWTC	5′-overhang	AWT–
ThaI	CG ↑ CG	Blunt	
TseI	G ↑ CWGC	5′-overhang	CWG–
TspRI	CAGTG(2/–7) –NNNNNNNCAGTGNN	3′-overhang	
Tth111I	GACN ↑ NNGTC	5′-overhang	N–
TthHB8I	T ↑ CGA	5′-overhang	CG–
Van91I	CCANNNN ↑ NTGG	3′-overhang	–NNN
VspI	AT ↑ TAAT	5′-overhang	TA–
XbaI	T ↑ CTAGA	5′-overhang	CTAG–
XcmI	CCANNNNN ↑ NNNNTGG	3′-overhang	–N
XcyI	C ↑ CCGGG	5′-overhang	CCGG–
XhoI	C ↑ TCGAG	5′-overhang	TCGA–
XhoII	R ↑ GATCY	5′-overhang	GATC–
XmaI	C ↑ CCGGG	5′-overhang	CCGG–
XmaIII	C ↑ GGCCG	5′-overhang	GGCC–
XmaCI	C ↑ CCGGG	5′-overhang	CCGG–
XmnI	GAANN ↑ NNTTC	Blunt	
XorII	CGAT ↑ CG	3′-overhang	–AT

[a] The arrow indicates the position of the cut site in the 5′→3′ strand in a palindromic recognition sequence. If the sequence is not palindromic then the position of the cut site is indicated by the numbers in brackets. The first number is the position of the cut site relative to the recognition sequence in the 5′→3′ strand and the second number is the position of the cut site relative to the recognition sequence in the 3′→5′ strand. A positive number indicates that the cut site lies downstream of the recognition sequence and a negative number indicates that it lies upstream. Conventional abbreviations are used for denoting non-unique nucleotides: R = G or A; Y = C or T; M = A or C; K = G or T; S = G or C; W = A or T; B = not A (C or G or T); D = not C (A or G or T); H = not G (A or C or T); V = not T (A or C or G); N = A or C or G or T.

[b] BcgI cuts both upstream and downstream of the recognition sequence, releasing the latter within a 32 bp fragment with 3′-overhangs of two nucleotides (–NN) at either end.

References

1. http://www.neb.com/rebase/rebase.html
2. Brown, T.A. (ed.) (1998). *Molecular biology labfax*, 2nd edn, Vol. 1. Academic Press, London.

Appendix 4
DNA and RNA modification enzymes

T. A. Brown

Department of Biomolecular Sciences, University of Manchester Institute of
Science and Technology, Manchester M60 IQD, UK

1 Introduction

Many different enzymes for DNA and RNA manipulations are now available
from commercial suppliers. A comprehensive list is provided in ref. 1 and most
are mentioned in various chapters of this book. The following is a summary.

2 DNA polymerases

DNA polymerases synthesize DNA from deoxyribonucleotide subunits.
Synthesis occurs in the 5′→3′ direction and is template-dependent with some
enzymes and template-independent with others. Most template-dependent DNA
polymerases also possess 5′→3′ and/or 3′→5′ exonuclease activities. Important
DNA polymerases are:

- DNA polymerase I of *E. coli* possesses a 5′→3′ DNA-dependent DNA polymerase activity as well as 5′→3′ and 3′→5′ exonuclease activities. It is used for
  DNA labelling by nick translation and for second-strand cDNA synthesis.

- Klenow polymerase is derived from DNA polymerase I by removal of the
  enzyme segment responsible for the 5′→3′ exonuclease. The enzyme is used
  for a number of applications, notably DNA labelling by random priming or
  end-filling, and conversion of 5′-overhangs to blunt ends.

- Sequenase is a modified version of the DNA polymerase of bacteriophage T7.
  It has no exonuclease activity and is ideal for DNA sequencing by the chain
  termination method.

- *Taq* DNA polymerase from *Thermus aquaticus* is heat-stable and so can be used
  in PCR and certain specialized DNA sequencing applications. A number of
  other thermostable DNA polymerases are also available and are used in various
  applications: these include *Pfu* DNA polymerase from *Pyrococcus furiosus* and
  Pwo DNA polymerase from *Pyrococcus woesei*.

- T4 DNA polymerase has a highly-active 3′→5′ exonuclease which enables it to carry out an end-replacement reaction that is useful in DNA labelling. It is also used to convert 3′-overhangs into blunt ends.

- Reverse transcriptases are obtained from various sources, including avian myeloblastosis virus (AMV) and Moloney murine leukaemia virus (M-MuLV). They are RNA-dependent DNA polymerases and so copy an RNA template into DNA, providing the basis for cDNA synthesis.

3 RNA polymerases

The most important RNA polymerases used in molecular biology research are three bacteriophage enzymes that recognize their own promoter sequences with high fidelity, and which can be used to synthesize large amounts of RNA in *in vitro* reactions. These three enzymes are the SP6, T3 and T7 RNA polymerases.

4 Nucleases

A great variety of nucleases are used to manipulate DNA and RNA molecules. Restriction endonucleases are described in *Appendix 3*. Other important nucleases are:

- Bal 31 nuclease possesses a complex set of activities that enable it to progressively shorten double-stranded blunt-ended molecules.

- S1 nuclease is single-strand specific and more active on DNA than RNA, so can be used to trim non-hybridized regions from DNA–RNA duplexes. Its use in S1 nuclease mapping allows the termini of transcripts to be located, along with intron boundaries.

- RNase H degrades the RNA component of a DNA–RNA hybrid and plays an important role in cDNA synthesis.

- DNase I is an endodeoxyribonuclease that has a number of important applications. It can be used to introduce nicks in double-stranded DNA molecules prior to labelling by nick translation, and can detect protein-binding sites in nuclease protection experiments.

- RNase A is an active ribonuclease that is used to remove RNA from DNA preparations.

5 Ligases

The T4 DNA ligase is usually used in construction of recombinant DNA molecules, but the *E. coli* ligase (which requires NAD rather than ATP as the cofactor) is also used. There is also a T4 RNA ligase.

6 End-modification enzymes

The ends of DNA and RNA molecules can be altered by treatment with an end-modification enzyme. Examples are:

- Alkaline phosphatase removes 5'-phosphate groups from single- and double-stranded DNA molecules, which prevents their self-ligation. Restricted cloning vectors are often treated with alkaline phosphatase. The resulting phosphatased ends cannot ligate to one another, but can ligate to the non-phosphatased ends of another DNA molecule. This means that recircularization of the cloning vector can occur only when new DNA is inserted, so only recombinant molecules are synthesized. BAP, from *E. coli*, and CIP have been used extensively in the past but both are relatively resistant to heat treatment and so it can be difficult to inactivate them after the phosphatase reaction has been carried out. More recently, heat-labile alkaline phosphatases from arctic bacteria and arctic shrimps have been introduced.

- T4 polynucleotide kinase adds phosphates to 5'-hydroxyl termini and is used in a DNA labelling procedure.

- Terminal deoxynucleotidyl transferase adds a single-stranded tail on to a (usually) blunt-ended molecule. This reaction has been used to add sticky ends to blunt-ended DNAs prior to construction of recombinant molecules.

Reference

1. Brown, T. A. (ed.) (1998). *Molecular biology labfax*, 2nd edn, Vol. 1. Academic Press, London.

Appendix 5
List of suppliers

Adams Healthcare, Lotherton Way, Garforth, Leeds LS25 2JY
Tel: 0044 1132 320066
Fax: 0044 1132 871317

Amersham Pharmacia Bio Tech
Pharmacia Biotech (Biochrom) Ltd., Unit 22, Cambridge Science Park, Milton Road, Cambridge CB4 0FJ, UK.
Tel: 01223 423723; Fax: 01223 420164
URL: http://www.biochrom.co.uk
Pharmacia and Upjohn Ltd., Davy Avenue, Knowlhill, Milton Keynes, Buckinghamshire MK5 8PH, UK
Tel: 01908 661101   Fax: 01908 690091
URL: http://www.eu.pnu.com

Anderman and Co. Ltd, 145 London Road, Kingston-upon-Thames, Surrey KT2 6NH, UK
Tel: 0181 541 0035   Fax: 0181 541 0623

Beckman Coulter Inc.
Beckman Coulter Inc., 4300 N Harbor Boulevard, PO Box 3100, Fullerton, CA 92834-3100, USA
Tel: 001 714 871 4848
Fax: 001 714 773 8283
URL: http://www.beckman.com
Beckman Coulter (UK) Ltd., Oakley Court, Kingsmead Business Park, London Road, High Wycombe, Buckinghamshire HP11 1JU, UK
Tel: 01494 441181   Fax: 01494 447558
URL: http://www.beckman.com

Becton Dickinson and Co.
Becton Dickinson and Co., 21 Between Towns Road, Cowley, Oxford OX4 3LY, UK
Tel: 01865 748844; Fax: 01865 781627
URL: http://www.bd.com
Becton Dickinson and Co., 1 Becton Drive, Franklin Lakes, NJ 07417-1883, USA
Tel: 001 201 847 6800
URL: http://www.bd.com

Bio 101 Inc.
Bio 101 Inc., c/o Anachem Ltd., Anachem House, 20 Charles Street, Luton, Bedfordshire LU2 0EB, UK
Tel: 01582 456666; Fax: 01582 391768
URL: http://www.anachem.co.uk
Bio 101 Inc., PO Box 2284, La Jolla, CA 92038-2284, USA
Tel: 001 760 598 7299
Fax: 001 760 598 0116
URL: http://www.bio101.com

Bio-Rad Laboratories Ltd
Bio-Rad Laboratories Ltd., Bio-Rad House, Maylands Avenue, Hemel Hempstead, Hertfordshire HP2 7TD, UK
Tel: 0181 328 2000   Fax: 0181 328 2550
URL: http://www.bio-rad.com
Bio-Rad Laboratories Ltd., Division Headquarters, 1000 Alfred Noble Drive, Hercules, CA 94547, USA
Tel: 001 510 724 7000
Fax: 001 510 741 5817
URL: http://www.bio-rad.com

BioWhittaker Molecular Applications
BioWhittaker Molecular Applications,
191 Thomaston Street, Rockland,
Maine 04841, USA
URL: http://www.bmaproducts.com
Supplied in the UK by Flowgen, Excelsior
Road, Ashby Park, Leicestershire LE65 1NG

Clontech
Clontech Laboratories UK Ltd, Unit 2, Intec
2, Wade Road, Basingstoke, Hants RG24
8NE, UK
Clontech Laboratories Inc, 1020 East
Meadow Circle, Palo Alto, CA 94303-4230,
USA
Clontech Laboratories GmbH, Tullastrasse 4,
D–69126 Heidelberg, Germany.
Clontech Laboratories Japan Inc, Shuwa Dai-
Ni Kayaba-Cho Bldg. 7F, 3-7-6 Nihonbashi
Kayabacho, Chuo-ku, Tokyo 130-0025,
Japan

CP Instrument Co. Ltd., PO Box 22, Bishop
Stortford, Hertfordshire CM23 3DX, UK
Tel: 01279 757711   Fax: 01279 755785
URL: http://www.cpinstrument.co.uk

Dupont
Dupont (UK) Ltd., Industrial Products
Division, Wedgwood Way, Stevenage,
Hertfordshire SG1 4QN, UK
Tel: 01438 734000   Fax: 01438 734382
URL: http://www.dupont.com
Dupont Co. (Biotechnology Systems
Division), PO Box 80024, Wilmington, DE
19880-002, USA
Tel: 001 302 774 1000
Fax: 001 302 774 7321
URL: http://www.dupont.com

Eastman Chemical Co., 100 North Eastman
Road, PO Box 511, Kingsport, TN 37662-
5075, USA
Tel: 001 423 229 2000
URL: http://www.eastman.com

Fisher Scientific
Fisher Scientific UK Ltd., Bishop Meadow
Road, Loughborough, Leicestershire LE11
5RG, UK
Tel: 01509 231166; Fax: 01509 231893
URL: http://www.fisher.co.uk
Fisher Scientific, Fisher Research, 2761
Walnut Avenue, Tustin, CA 92780, USA
Tel: 001 714 669 4600
Fax: 001 714 669 1613
URL: http://www.fishersci.com

Flowgen, Novara House, Excelsior Road,
Ashby Park, Ashby de la Zouch,
Leicestershire LE65 1NG
Tel: 0870 6000152
Fax: 0044 1530 4192950
URL: www.flowgen.co.uk

Fluka
Fluka, PO Box 2060, Milwaukee, WI 53201,
USA
Tel: 001 414 273 5013
Fax: 001 414 2734979
URL: http://www.sigma-aldrich.com
Fluka Chemical Co. Ltd., PO Box 260,
CH–9471, Buchs, Switzerland.
Tel: 0041 81 745 2828
Fax: 0041 81 756 5449
URL: http://www.sigma-aldrich.com

Hybaid
Hybaid Ltd., Action Court, Ashford Road,
Ashford, Middlesex TW15 1XB, UK
Tel: 01784 425000   Fax: 01784 248085
URL: http://www.hybaid.com
Hybaid US, 8 East Forge Parkway, Franklin,
MA 02038, USA.
Tel: 001 508 541 6918
Fax: 001 508 541 3041
URL: http://www.hybaid.com

HyClone Laboratories, 1725 South HyClone
Road, Logan, UT 84321, USA
Tel: 001 435 753 4584
Fax: 001 435 753 4589
URL: http://www.hyclone.com

Invitrogen
Invitrogen BV, PO Box 2312, CH–9704
Groningen, The Netherlands
Tel: 00800 5345 5345
Fax: 00800 7890 7890
URL: http://www.invitrogen.com
Invitrogen Corp., 1600 Faraday Avenue,
Carlsbad, CA 92008, USA
Tel: 001 760 603 7200
Fax: 001 760 603 7201
URL: http://www.invitrogen.com

Life Technologies
Life Technologies Ltd., PO Box 35, Free
Fountain Drive, Inchinnan Business Park,
Paisley PA4 9RF, UK
Tel: 0800 269210   Fax: 0800 838380
URL: http://www.lifetech.com
Life Technologies Inc., 9800 Medical Center
Drive, Rockville, MD 20850, USA
Tel: 001 301 610 8000
URL: http://www.lifetech.com

Merck Sharp & Dohme
Merck Sharp & Dohme Research
Laboratories, Neuroscience Research Centre,
Terlings Park, Harlow, Essex CM20 2QR, UK
Web site: http://www.msd-nrc.co.uk
MSD Sharp and Dohme GmbH, Lindenplatz
1, D–85540, Haar, Germany.
Web site: http://www.msd-deutschland.com

Millipore
Millipore (UK) Ltd., The Boulevard,
Blackmoor Lane, Watford, Hertfordshire
WD1 8YW, UK
Tel: 01923 816375   Fax: 01923 818297
URL: http://www.millipore.com/local/UK.htm
Millipore Corp., 80 Ashby Road, Bedford,
MA 01730, USA
Tel: 001 800 645 5476
Fax: 001 800 645 5439
URL: http://www.millipore.com

Molecular Probes
Molecular Probes, 4849 Pitchford Avenue,
Eugene, Oregon 97402, USA

Molecular Research Center, 5645
Montgomery Road, Cincinnati, OH 45212
Tel: 001 513 841 0080
Fax: 001 513 841 0090

New England Biolabs, 32 Tozer Road,
Beverley, MA 01915-5510, USA
Tel: 001 978 927 5054

Nikon
Nikon Corp., Fuji Building, 2-3, 3-chome,
Marunouchi, Chiyoda-ku, Tokyo 100, Japan
Tel: 00813 3214 5311; Fax: 00813 3201 5856
URL: http://www.nikon.co.jp/main/index_e.htm
Nikon Inc., 1300 Walt Whitman Road,
Melville, NY 11747-3064, USA
Tel: 001 516 547 4200
Fax: 001 516 547 0299
URL: http://www.nikonusa.com

Nycomed
Nycomed Amersham plc, Amersham Place,
Little Chalfont, Buckinghamshire HP7 9NA,
UK
Tel: 01494 544000   Fax: 01494 542266
URL: http://www.amersham.co.uk
Nycomed Amersham, 101 Carnegie Center,
Princeton, NJ 08540, USA.
Tel: 001 609 514 6000
URL: http://www.amersham.co.uk

Perkin Elmer Ltd, Post Office Lane,
Beaconsfield, Buckinghamshire HP9 1QA,
UK
Tel: 01494 676161
URL: http://www.perkin-elmer.com

Pharmacia (please see Amersham Pharmacia
Bio Tech)

Philip Harris, 618 Western Avenue, Park
Royal, London W3 0TE
Tel: 0044 181 992 5555
Fax: 0044 181 993 8020

Promega
Promega UK Ltd., Delta House, Chilworth
Research Centre, Southampton SO16 7NS, UK
Tel: 0800 378994; Fax: 0800 181037

URL: http://www.promega.com
Promega Corp., 2800 Woods Hollow Road,
Madison, WI 53711-5399, USA
Tel: 001 608 274 4330
Fax: 001 608 277 2516
URL: http://www.promega.com

Qiagen
Qiagen UK Ltd., Boundary Court, Gatwick
Road, Crawley, West Sussex RH10 2AX, UK
Tel: 01293 422911; Fax: 01293 422922
URL: http://www.qiagen.com
Qiagen Inc., 28159 Avenue Stanford,
Valencia, CA 91355, USA
Tel: 001 800 426 8157
Fax: 001 800 718 2056
URL: http://www.qiagen.com

Roche Diagnostics
Roche Diagnostics Ltd., Bell Lane, Lewes,
East Sussex BN7 1LG, UK
Tel: 01273 484644   Fax: 01273 480266
URL: http://www.roche.com
Roche Diagnostics Corp., 9115 Hague Road,
PO Box 50457, Indianapolis, IN 46256, USA
Tel: 001 317 845 2358
Fax: 001 317 576 2126
URL: http://www.roche.com
Roche Diagnostics GmbH, Sandhoferstrasse
116, 68305 Mannheim, Germany
Tel: 0049 621 759 4747
Fax: 0049 621 759 4002
URL: http://www.roche.com

Schleicher and Schuell Inc., Keene, NH
03431A, USA   Tel: 001 603 357 2398

Shandon Scientific Ltd, 93–96 Chadwick
Road, Astmoor, Runcorn, Cheshire WA7
1PR, UK   Tel: 01928 566611
URL: http://www.shandon.com

Sigma–Aldrich
Sigma–Aldrich Co. Ltd., The Old Brickyard,
New Road, Gillingham, Dorset XP8 4XT, UK
Tel: 01747 822211; Fax: 01747 823779
URL: http://www.sigma-aldrich.com
Sigma-Aldrich Co. Ltd., Fancy Road, Poole,
Dorset BH12 4QH, UK
Tel: 01202 722114; Fax: 01202 715460
URL: http://www.sigma-aldrich.com

Sigma Chemical Co., PO Box 14508, St Louis,
MO 63178, USA.
Tel: 001 314 771 5765
Fax: 001 314 771 5757
URL: http://www.sigma-aldrich.com

Stratagene
Stratagene Europe, Gebouw California,
Hogehilweg 15, 1101 CB Amsterdam
Zuidoost, The Netherlands
Tel: 00800 9100 9100
URL: http://www.stratagene.com
Stratagene Inc., 11011 North Torrey Pines
Road, La Jolla, CA 92037, USA.
Tel: 001 858 535 5400
URL: http://www.stratagene.com

United States Biochemical, PO Box 22400,
Cleveland, OH 44122, USA
Tel: 001 216 464 9277

Index

Page numbers in italic signify references to figures. Page numbers in bold denote entries in tables.